NATIONAL COOPERATIVE HIGHWAY RESEARCH PROGRAM

NCHRP SYNTHESIS 346

State Construction Quality Assurance Programs

A Synthesis of Highway Practice

CONSULTANT
CHARLES S. HUGHES
Charlottesville, Virginia

TOPIC PANEL

JAMES BURATI, *Clemson University*
GEORGENE GEARY, *Georgia Department of Transportation*
CAL J. GENDREAU, *North Dakota Department of Transportation*
FREDERICK HEJL, *Transportation Research Board*
GENE MALLETTE, *California Department of Transportation*
LLOYD M. WELKER, *Ohio Department of Transportation*
RICHARD O. WOLTERS, *Minnesota Asphalt Paving Association*
PETER A. KOPAC, *Federal Highway Administration* (Liaison)
MICHAEL RAFALOWSKI, *Federal Highway Administration* (Liaison)

SUBJECT AREAS
Pavement Design, Management, and Performance; Bridges, Other Structures, Hydraulics, and Hydrology; Maintenance

Research Sponsored by the American Association of State Highway and Transportation Officials in Cooperation with the Federal Highway Administration

TRANSPORTATION RESEARCH BOARD

WASHINGTON, D.C.
2005
www.TRB.org

NATIONAL COOPERATIVE HIGHWAY RESEARCH PROGRAM

Systematic, well-designed research provides the most effective approach to the solution of many problems facing highway administrators and engineers. Often, highway problems are of local interest and can best be studied by highway departments individually or in cooperation with their state universities and others. However, the accelerating growth of highway transportation develops increasingly complex problems of wide interest to highway authorities. These problems are best studied through a coordinated program of cooperative research.

In recognition of these needs, the highway administrators of the American Association of State Highway and Transportation Officials initiated in 1962 an objective national highway research program employing modern scientific techniques. This program is supported on a continuing basis by funds from participating member states of the Association and it receives the full cooperation and support of the Federal Highway Administration, United States Department of Transportation.

The Transportation Research Board of the National Academies was requested by the Association to administer the research program because of the Board's recognized objectivity and understanding of modern research practices. The Board is uniquely suited for this purpose as it maintains an extensive committee structure from which authorities on any highway transportation subject may be drawn; it possesses avenues of communications and cooperation with federal, state, and local governmental agencies, universities, and industry; its relationship to the National Research Council is an insurance of objectivity; it maintains a full-time research correlation staff of specialists in highway transportation matters to bring the findings of research directly to those who are in a position to use them.

The program is developed on the basis of research needs identified by chief administrators of the highway and transportation departments and by committees of AASHTO. Each year, specific areas of research needs to be included in the program are proposed to the National Research Council and the Board by the American Association of State Highway and Transportation Officials. Research projects to fulfill these needs are defined by the Board, and qualified research agencies are selected from those that have submitted proposals. Administration and surveillance of research contracts are the responsibilities of the National Research Council and the Transportation Research Board.

The needs for highway research are many, and the National Cooperative Highway Research Program can make significant contributions to the solution of highway transportation problems of mutual concern to many responsible groups. The program, however, is intended to complement rather than to substitute for or duplicate other highway research programs.

NCHRP SYNTHESIS 346

Project 20-5 FY 2003 (Topic 35-01)
ISSN 0547-5570
ISBN 0-309-09749-5
Library of Congress Control No. 2005926684

Price $17.00

NOTICE

The project that is the subject of this report was a part of the National Cooperative Highway Research Program conducted by the Transportation Research Board with the approval of the Governing Board of the National Research Council. Such approval reflects the Governing Board's judgment that the program concerned is of national importance and appropriate with respect to both the purposes and resources of the National Research Council.

The members of the technical committee selected to monitor this project and to review this report were chosen for recognized scholarly competence and with due consideration for the balance of disciplines appropriate to the project. The opinions and conclusions expressed or implied are those of the research agency that performed the research, and, while they have been accepted as appropriate by the technical committee, they are not necessarily those of the Transportation Research Board, the National Research Council, the American Association of State Highway and Transportation Officials, or the Federal Highway Administration, U.S. Department of Transportation.

Each report is reviewed and accepted for publication by the technical committee according to procedures established and monitored by the Transportation Research Board Executive Committee and the Governing Board of the National Research Council.

Published reports of the

NATIONAL COOPERATIVE HIGHWAY RESEARCH PROGRAM

are available from:

Transportation Research Board
Business Office
500 Fifth Street, NW
Washington, DC 20001

and can be ordered through the Internet at:

http://www.national-academies.org/trb/bookstore

Printed in the United States of America

THE NATIONAL ACADEMIES

Advisers to the Nation on Science, Engineering, and Medicine

The **National Academy of Sciences** is a private, nonprofit, self-perpetuating society of distinguished scholars engaged in scientific and engineering research, dedicated to the furtherance of science and technology and to their use for the general welfare. On the authority of the charter granted to it by the Congress in 1863, the Academy has a mandate that requires it to advise the federal government on scientific and technical matters. Dr. Bruce M. Alberts is president of the National Academy of Sciences.

The **National Academy of Engineering** was established in 1964, under the charter of the National Academy of Sciences, as a parallel organization of outstanding engineers. It is autonomous in its administration and in the selection of its members, sharing with the National Academy of Sciences the responsibility for advising the federal government. The National Academy of Engineering also sponsors engineering programs aimed at meeting national needs, encourages education and research, and recognizes the superior achievements of engineers. Dr. William A. Wulf is president of the National Academy of Engineering.

The **Institute of Medicine** was established in 1970 by the National Academy of Sciences to secure the services of eminent members of appropriate professions in the examination of policy matters pertaining to the health of the public. The Institute acts under the responsibility given to the National Academy of Sciences by its congressional charter to be an adviser to the federal government and, on its own initiative, to identify issues of medical care, research, and education. Dr. Harvey V. Fineberg is president of the Institute of Medicine.

The **National Research Council** was organized by the National Academy of Sciences in 1916 to associate the broad community of science and technology with the Academy's purposes of furthering knowledge and advising the federal government. Functioning in accordance with general policies determined by the Academy, the Council has become the principal operating agency of both the National Academy of Sciences and the National Academy of Engineering in providing services to the government, the public, and the scientific and engineering communities. The Council is administered jointly by both the Academies and the Institute of Medicine. Dr. Bruce M. Alberts and Dr. William A. Wulf are chair and vice chair, respectively, of the National Research Council.

The **Transportation Research Board** is a division of the National Research Council, which serves the National Academy of Sciences and the National Academy of Engineering. The Board's mission is to promote innovation and progress in transportation through research. In an objective and interdisciplinary setting, the Board facilitates the sharing of information on transportation practice and policy by researchers and practitioners; stimulates research and offers research management services that promote technical excellence; provides expert advice on transportation policy and programs; and disseminates research results broadly and encourages their implementation. The Board's varied activities annually engage more than 5,000 engineers, scientists, and other transportation researchers and practitioners from the public and private sectors and academia, all of whom contribute their expertise in the public interest. The program is supported by state transportation departments, federal agencies including the component administrations of the U.S. Department of Transportation, and other organizations and individuals interested in the development of transportation. **www.TRB.org**

www.national-academies.org

FOREWORD

By Staff
Transportation
Research Board

Highway administrators, engineers, and researchers often face problems for which information already exists, either in documented form or as undocumented experience and practice. This information may be fragmented, scattered, and unevaluated. As a consequence, full knowledge of what has been learned about a problem may not be brought to bear on its solution. Costly research findings may go unused, valuable experience may be overlooked, and due consideration may not be given to recommended practices for solving or alleviating the problem.

There is information on nearly every subject of concern to highway administrators and engineers. Much of it derives from research or from the work of practitioners faced with problems in their day-to-day work. To provide a systematic means for assembling and evaluating such useful information and to make it available to the entire highway community, the American Association of State Highway and Transportation Officials—through the mechanism of the National Cooperative Highway Research Program—authorized the Transportation Research Board to undertake a continuing study. This study, NCHRP Project 20-5, "Synthesis of Information Related to Highway Problems," searches out and synthesizes useful knowledge from all available sources and prepares concise, documented reports on specific topics. Reports from this endeavor constitute an NCHRP report series, *Synthesis of Highway Practice.*

This synthesis series reports on current knowledge and practice, in a compact format, without the detailed directions usually found in handbooks or design manuals. Each report in the series provides a compendium of the best knowledge available on those measures found to be the most successful in resolving specific problems.

PREFACE

This synthesis describes the current quality assurance (QA) practices of state and federal departments of transportation with regard to highway materials and construction. The report focuses on the strategies and practices used by agencies to ensure quality. Because QA is viewed differently among the agencies, methods and procedures that constitute the QA programs of highway agencies also differ significantly. This synthesis summarizes these methods and procedures to the greatest extent feasible, including information on quality control, acceptance, independent assurance, and training/certification. It includes discussion of statistically based specifications, QA specifications, FHWA QA procedures for construction (complying with 23 CFR 637), performance-related specifications, optimal procedures for QA specifications, the use of consultants, and resource allocation.

This synthesis report of the Transportation Research Board contains information developed from a literature review of QA practices. The results of a survey questionnaire that detail the current state of the practice of state, federal, and Canadian QA programs supplement the literature review. To better understand the terms used in this synthesis, terms related to QA programs and specification are defined, as adapted from TRB's *Glossary of Highway Quality Assurance Terms.*

A panel of experts in the subject area guided the work of organizing and evaluating the collected data and reviewed the final synthesis report. A consultant was engaged to collect and synthesize the information and to write the report. Both the consultant and the members of the oversight panel are acknowledged on the title page. This synthesis is an immediately useful document that records the practices that were acceptable within the limitations of the knowledge available at the time of its preparation. As progress in research and practice continues, new knowledge will be added to that now at hand.

CONTENTS

STATE CONSTRUCTION QUALITY ASSURANCE PROGRAMS

SUMMARY

This synthesis describes the current quality assurance (QA) practices of state and federal departments of transportation with regard to highway materials and construction. The background information for the report was developed from a review of the literature of QA practices. The results of a survey questionnaire that detail the current state of the practice of state, federal, and Canadian QA programs supplement the literature review.

Departments of transportation have come to realize the importance of QA from the experience that failure to conform to either material or construction specifications can result in the failure of highway components. Construction QA programs are intended to ensure that the quality of the materials and construction incorporated in highway products is satisfactory; thus, such programs are many faceted. The use of QA programs has evolved since the 1960s into what are now, sometimes, second or third generation QA programs. QA programs contain three main ingredients, quality control (QC), acceptance, and independent assurance (IA). The manner in which these ingredients are administered and blended makes for many different versions of such programs. To provide some scale of the magnitude of the differences that are found in these programs, examples of decisions agencies must consider are the following:

- Choosing the attributes to use for QC and for acceptance.
- Choosing the test methods to use for QC and for acceptance.
- Deciding on the point of sampling to use for QC and for acceptance.
- Deciding who establishes the frequency for QC tests.
- Deciding how to establish the QC tests.
- Deciding on the quality measure to use for acceptance.
- Deciding whether accept/reject or pay adjustment provisions will be used.
- Deciding what levels of risks are appropriate for the agency and contractor.
- Deciding whether contractor tests will be used in the acceptance decision:
  - if they are, deciding on the type of verification system that will be used;
  - whether the agency will use split samples, independent samples, or both; and
  - the purpose of the verification.
- Deciding if training and/or certification will be required. And, if required, determining who will do it.
- Deciding how the IA function will be administered.

The ways these issues have been addressed reflect the evolutionary process that QA programs have undergone over the last 30 years. Some of the major changes that have taken place emerged from Title 23, Part 637, Code of Federal Regulations (23 CFR 637), the FHWA's *Quality Assurance Procedures for Construction.* This regulation was adopted in 1995 and requires that each state highway agency develop a QA program for the National Highway System. The program is structured to ensure that the materials and workmanship incorporated into each federal-aid highway construction project on the National Highway System are in conformity with the requirements of the approved plans and specifications, including approved changes. The responses to the questionnaire indicated that the evolution is continuing.

Currently, the strategies and practices used by state and federal highway agencies to ensure quality employ a wide variety of QA approaches to meet 23 CFR 637. The questionnaire responses indicated that the type of QA program varies not only among, but also within, agencies depending on the material and construction area specified. This creates what can be considered a spectrum of QA programs. At one end are QA programs that rely primarily on materials and methods provisions. At the other end of the spectrum are QA programs in which agencies use contractor test results as part of the acceptance decision. In between are various combinations of QC and acceptance provisions where the agency assumes a greater or lesser role in QC, leaving the complementary lesser or greater role for the contractor. Agencies tend to use materials and methods provisions to a greater extent for soils and embankment specifications than for other materials and construction and to use contractor test results in the acceptance decision more often for hot-mix asphalt.

Most agencies require contractor QC for at least one material, and several require it for the majority of materials. Many agencies retain the entire acceptance function; however, the number of agencies using contractor test results in the acceptance decision is increasing. Because QC and acceptance are performed for two different purposes, it is desirable to separate these functions. This separation is often not clear. The third QA function, IA, is being conducted by all agencies in compliance with 23 CFR 637; however, the manner in which IA is organized within an agency varies greatly, as does the level of staffing, even when normalized by agency budgets.

When using contractor test results in the acceptance decision, 23 CFR 637 requires that verification testing be done by the agency. The type of verification currently being used varies greatly from agency to agency. Some agencies use a stronger statistical verification system than others. A considerable number of agencies use a weaker verification system that is less sensitive to differences between agency and contractor test results.

The use of consultants is widespread as a means to comply with 23 CFR 637 while coping with personnel reductions. More than 75% of the responding agencies stated that they use consultants. Most use the consultants in place of or as a supplement to agency acceptance testing, and a significant number use them in place of or as a supplement to contractor QC testing.

The use of innovative practices is not widespread. Eight agencies use warranties on a routine basis and seven routinely use design–build. No other innovative practices were found to be used to an appreciable extent.

Changes in QA programs are anticipated by many agencies. The products where changes are expected vary from entire pavement QA programs to individual material and construction components.

CHAPTER ONE

INTRODUCTION

State and federal departments of transportation realize the importance of quality assurance (QA). This realization has come from the experience that failure to conform to either material or construction specifications can result in the premature failure of highway components. Construction QA programs are intended to ensure that the quality of the materials and construction incorporated in the highway products is satisfactory. Therefore, such programs are many faceted.

Since the 1960s, QA programs have evolved into what are sometimes now second or third generation QA programs. This evolutionary process is reflected in the changes allowed in Title 23, Part 637, Code of Federal Regulations (23 CFR 637), the FHWA's *Quality Assurance Procedures for Construction* (*1*). This regulation was adopted in 1995 and requires each state highway agency (SHA) to develop a QA program for the National Highway System (NHS). The program is designed to ensure that the materials and workmanship incorporated into each federal-aid highway construction project on the NHS are in conformity with the requirements of the approved plans and specifications, including approved changes.

Currently, the strategies and practices used by state and federal highway agencies to ensure quality employ a wide variety of QA approaches to meet the regulations as revised under 23 CFR 637. This synthesis is intended to bring the information on these varied approaches into a single resource document. These approaches include both statistical QA and nonstatistical QA.

PURPOSE AND SCOPE

This synthesis of the current practice of QA programs was undertaken to describe the wide range of methods and procedures that agencies use to ensure quality. The implementation of 23 CFR 637 has changed the way many agencies approach the measure of quality. This changing environment has created a need to document the various approaches and highlight new and innovative strategies used by agencies.

This synthesis summarizes the current methods and procedures that constitute the QA programs of highway agencies. The objective is to document how agencies conduct their QA programs. Therefore, the scope of the report is on the strategies and practices employed by agencies to ensure quality. Because QA is viewed quite differently among the agencies, the QA methods and procedures, logically, also differ significantly among agencies. The form of specification often varies within an agency depending on the material or construction item. It is intended that this synthesis capture these methods and procedures to the greatest extent feasible.

This synthesis explores the agencies' QA programs and their future directions for these programs. Because there are many important parts of a QA program, not all are covered in detail. For instance, the many facets of dispute resolution, detailed information on quality control (QC) charts, and pay adjustment forms are beyond the scope of this synthesis. However, rudimentary information on both QC and pay adjustment are included. Both portland cement concrete paving (PCCP) and PCC structures are included, because agencies often have different types of specifications for each. Other structural materials, such as steel, timber, and prestressed and precast concrete, are not specifically included, but are discussed in a recent *Research Results Digest* (*2*).

BACKGROUND

The genesis of the evolution in QA programs lies in the AASHO Road Test (1956–1958) and the analysis that emanated from this historic study. Before the AASHO Road Test, specifications, with few exceptions, were materials and methods specifications. "It was during the construction of this project [the AASHO Road Test] that a sufficient number of unbiased test results of construction materials and techniques became available to expose the true variability of these results and their relationship to specifications" (*3*). The analysis of the results of the material and construction properties from this test road revealed variabilities that were much greater than expected.

The evolution that has taken place in QA programs during the intervening years has produced several forms of specifications (*4*). This evolution has been driven by several factors, two of which are the previously mentioned AASHO Road Test that showed the importance of recognizing variability in the specifications and the construction of the Interstate highway system. This large road-building endeavor encouraged technological advances that increased construction speed. "State highway agencies may have been at least partly motivated to implement QA specifications because they had too few inspectors to oversee the rapidly growing interstate system under method specification" (*5*).

Statistically Based Specifications

Soon after the results of the AASHO Road Test were published, many agencies started measuring the variability of typical material and construction properties as a first step in establishing specification limits for statistically based specifications. Because these types of specifications were being used for the first time, a great deal of education in the proper use of statistical tools was necessary. These types of specifications, developed during the 1960s, were for the most part what are called "Variability Known" or "Variability Assumed" specifications. Such specifications concentrated on controlling the average of the product or process.

At about the same time as the implementation of statistically based specifications, the use of disincentive pay factors, often called penalties or negative price adjustments, was initiated for products that did not meet the specifications. These adjusted payments were used instead of product removable (accept/reject) or "shut-downs" of the operation. These pay factors were not viewed favorably by the private sector. Although the decision to use disincentive clauses in statistically based specifications was independent of the use of the statistical tools, contractors related the disincentives to the use of statistics and viewed the disincentives as being unduly harsh (*6*).

QA Specifications

By the 1970s, the statistically based specifications had been incorporated into QA programs with a strong dependence on statistical analysis (*7*). With the development of these programs came the recognition of a need for separate quality (process) control and acceptance functions. Part of this recognition was the realization by the specifying agency that the contractor, or producer, was in the best position to conduct the process control function, because it depended on the contractor's personnel and equipment. The acceptance function was generally agreed to be an agency function to ensure that "satisfactory quality control has been exercised and that the proper degree of compliance to the specifications has been attained" (*7*). These definitions of the parts of a QA system have been formalized and adopted in the *AASHTO Quality Assurance Guide Specification* (*8*).

23 CFR 637

Starting in the 1980s, SHAs began taking a critical look at testing personnel assigned to contractors' facilities. In many cases, the contractor had assumed the testing and inspection activities associated with QC; therefore, there was a growing perception that a duplication of testing was taking place: QC testing by the contractor and acceptance testing by the agency. In some states, this was a primary reason for making the decision to remove the agency inspector/technician from the contractor's facility, and was coupled with the emphasis on reducing the number of government personnel. This created new management techniques heretofore unthought of (*9*). Because the contractor was doing inspection and performing testing more frequently than the agency, the question was asked, why not use the contractor's test results in the acceptance decision? Although this could be done on state-funded construction at that time, federal regulation did not allow for this type of procedure. Consequently, in the early 1990s, the FHWA decided to review regulations that would allow for such test results to be used in the acceptance decision. An unpublished report, *Limits of the Use of Contractor Performed Sampling and Testing*, recommended that contractor sampling and testing be used in acceptance programs (*1*). In 1995, 23 CFR 637 was adopted, which implemented this recommendation.

The full text and commentary of 23 CFR 637 on the final ruling is included in Appendix A. Gary Smith, in *NCHRP Synthesis of Highway Practice 263*, stated that "The regulation opens new avenues for innovative materials and construction acceptance procedures. The regulation enables transportation agencies to incorporate contractor test data into their quality acceptance procedures, and specifies laboratory certification requirements and personnel qualifications" (*9*, p. 3) However, the contractor test data can only be used "provided adequate checks and balances are in place to protect the public investment" (*1*).

Performance-Related Specifications

The evolution has continued to where performance-related specifications (PRS) are now being developed. In 1995, the topic of PRS was addressed in *NCHRP Synthesis of Highway Practice 212* (*10*). At that time, PRS were in their infancy and used infrequently. Prototype PRS now are available for both hot-mix asphalt (HMA) and PCC pavements (*11*).

Optimal Procedures for QA Specifications

A recent publication, *Optimal Procedures for Quality Assurance Specifications* (OPQAS), was written under the auspices of the FHWA. This publication is intended to serve as a how-to manual for agencies interested in writing a new or modifying an existing QA specification (*12*). It is seen as part of the evolutionary process because of the cutting edge discussion of risk and risk analysis for acceptance plans.

DEFINITIONS

One problem associated with QA programs and specifications since their inception has been differing interpretations of the specialized vocabulary used in these programs. To better understand the terms used in this synthesis, the terms related to QA programs and specifications are defined here.

These terms have been adapted from the TRB Transportation Circular, *Glossary of Highway Quality Assurance Terms* (*13*).

Acceptance—sampling and testing, or inspection, to determine the degree of compliance with contract requirements.

End-result specifications—specifications that require the contractor to take the entire responsibility for supplying a product or an item of construction. The highway agency's responsibility is to either accept or reject the final product or to apply a price adjustment commensurate with the degree of compliance with the specifications.

Independence assurance (IA)—management tool that requires a third party, not directly responsible for process control or acceptance, to provide an independent assessment of the product and/or the reliability of test results obtained from the process control and acceptance testing. (The results of IA tests are not to be used as the basis of product acceptance.) This definition differs from that of 25 CFR 637, which defines IA programs as "activities that are an unbiased and independent evaluation of all sampling and testing procedures used in the acceptance program."

Lot (also called population)—specific quantity of similar material, construction, or units of product subjected to either an acceptance or process control decision. (A lot, as a whole, is assumed to be produced by the same process.)

Materials and methods specifications (also called method specifications, recipe specifications, or prescriptive specifications)—specifications that direct the contractor to use specified materials in definite proportions and specific types of equipment and methods to place the material. Each step is directed by a representative of the highway agency.

Performance-related specifications—QA specifications that describe levels of key materials and construction quality characteristics that have been found to correlate with fundamental engineering properties that predict performance. These characteristics (e.g., air voids in asphalt concrete and compressive strength of PCC) are amenable to acceptance testing at the time of construction.

Performance specifications—specifications that describe how the finished product should perform over time.

Quality assurance (QA)—all planned and systematic actions necessary to provide confidence that a product or facility will perform satisfactorily in service. (This broad definition involves more activities than are covered in this synthesis; however, the term is defined to provide a basis of reference.)

Quality assurance specifications—combination of end-result specifications and materials and methods specifications. The contractor is responsible for QC (process control), and the highway agency is responsible for acceptance of the product. (QA specifications typically are statistically based specifications that use methods, such as random sampling and lot-by-lot testing, which let the contractor know if the operations are producing an acceptable product.)

Quality control (QC) (also called process control)—QA actions and considerations necessary to *assess and adjust* production and construction processes so as to *control* the level of quality being produced in the end product (emphasis added).

Statistically based specifications (also called statistical specifications or statistically oriented specifications)—specifications based on random sampling, and in which properties of the desired product or construction are described by appropriate statistical parameters.

Verification—process of determining or testing the truth or accuracy of test results by examining the data and/or providing objective evidence. [Verification sampling and testing may be part of an independent assurance program (to verify contractor QC testing or agency acceptance) or part of an acceptance program (to verify contractor testing used in the agency's acceptance decision).] This definition differs from that in 25 CFR 637, which defines verification sampling and testing as "sampling and testing performed to validate the quality of the product."

GENERAL SURVEY QUESTIONNAIRE INFORMATION

A survey was conducted to solicit information on the QA methods and procedures used by government agencies. (A summary of the responses to the survey questionnaire is in Appendix B.) The survey questionnaire was sent to the 50 SHAs, the District of Columbia, FHWA Federal Lands Division, and Canadian provinces. Responses were received from 43 SHAs, the District of Columbia, and the FHWA Federal Lands Division. Although the Canadian provinces have no requirements under 23 CFR 637, five were sufficiently interested in the subject of QA programs to complete the questionnaire. (A list of the respondents is in Appendix C.) The QA practices of the Canadian provinces as obtained from the responses are briefly summarized under each material/construction area, but are not included in the general discussion of each material/construction area. The questionnaire was organized into eight parts to obtain information related to

- Soils and Embankments,
- Aggregate Base and Subbase,
- Hot-Mix Asphalt,
- PCC Paving,
- Structural PCC,
- Independent Assurance,
- Use of Consultants and/or Innovative Practices, and
- Future of QA Programs.

It will be noticed in the discussion of each material/construction area that the numbers cited are often greater than the number of responses. This is because many of the questions were not "attribute specific," and therefore more than one procedure could be used in a particular material/construction area. For example, in Soils and Embankments, the attributes used most often for both QC and acceptance are moisture content and compaction. An agency could use primarily material and method provisions with an accept/reject plan for moisture content and also use contractor test results as part of the acceptance decision for compaction.

Another apparent anomaly may be found in the area of training and certification. Question eight in each material/construction area asked if the agency *required* training or certification under a particular program. Several respondents replied by checking more than one program, leading to the interpretation of (for the purposes of this synthesis) these responses as *using* or *allowing* training or certification under these programs, not requiring it.

CHAPTER TWO

QUALITY ASSURANCE PROGRAMS

The development of QA programs has been an evolutionary process and the forms and ingredients of QA programs vary appreciably from agency to agency. By definition, QA specifications combine end-result and materials and methods requirements. However, the way they are combined and the emphasis on each leads to the diversity. The nature of the materials and construction also affects the diversity in QA programs. For example, some agencies have found that because of the relatively high heterogeneity of in-place soils and embankments it is often more difficult to use statistically based specifications for these materials than for plant-produced materials and therefore they rely more heavily on materials and methods specifications. The initiation of 23 CFR 637 further affected the diversity in QA programs, because, although the rule provides more flexibility to the agency, there are important requirements if contractor test results are used in the acceptance decision. Specifically, the preamble of the final rule making of 23 CFR 637 states that:

> The overall intent of the program is to provide adequate assurance that the public is receiving the desired quality in the product produced by the contractor. The first level of assurance is provided by qualified laboratories and testing personnel. This assures that the equipment and personnel are capable of performing the tests properly. The second level of assurance is by the IA program. This level assures that the testers and equipment remain capable of performing the tests properly. The third level of assurance is provided by verification sampling and testing. This level assures the quality of the product (*1*).

In general, irrespective of what party to the contract performs the procedures, there are three integral parts that are necessary in an effective QA program: QC, acceptance, and IA. Agencies differ in the way they conduct these functions, not only from agency to agency, but within an agency, depending on the type of specification and material or construction area.

As part of the evolution of QA, the three functions have likewise evolved. Before 1960, little thought was given to any function except inspection and sampling and testing to determine if the specification limits were being met. The specifications used at that time were typically materials and methods specifications and, as such, sampling and testing were done by the specifying agency. Often, if the specification was not met, it was assumed that the test result was in error, and the material or construction was sampled and tested again. This assumption was made primarily because the agency was making the decisions concerning the contractor's operation, and the concept of variability was not well understood. At that time, there was no formal QC, and before 1962, no formal IA function (*3,7*). Following the AASHO Road Test analysis and a report from the Congressional House Committee on Oversights and Investigations ("Blatnik Committee"), agencies sought better ways to determine if specifications were being met (*3,7*).

NCHRP Synthesis of Highway Practice 38: *Statistically Oriented End-Result Specifications* (*3*) offers the following commentary:

> In 1963 the Bureau of Public Roads obtained under contract a report entitled "A Plan for Expediting the Use of Statistical Concepts in Highway Acceptance Specifications" [*6* (*14*)]. Based on this report the bureau circulated to the state highway agencies in 1965 a publication entitled "The Statistical Approach to Quality Control in Highway Construction" [*7* (*15*)]. This booklet contains explicit instructions for measuring current quality in a statistically valid manner and for determining the proportions of the total variability of the quality measurements due to actual variability in the materials or variability caused by sampling or testing. Programs for most types of highway materials and construction are included as well as a computer program for determining the statistical parameters from the test values. Also in 1965, the National Cooperative Highway Research Program (NCHRP) issued *NCHRP Report 17,* "Development of Guidelines for Practical and Realistic Construction Specifications" [*8* (*16*)] (*3*, p. 3).

The outcome of these studies and investigations was the requirement to conduct IA sampling and testing on FHWA-funded projects and was the beginning of formal QC initiatives.

The intended function of each part of QA is important because each function should supplement the other. The analogy has been used of QA being similar to a three-legged stool, with one leg being QC, one leg being acceptance, and the third leg being IA (*17*). With any leg missing, the whole is unbalanced. The present-day concept of QA is that QC is the responsibility of the contractor, acceptance is the responsibility of the agency (although this responsibility may involve contractor test results), and IA is conducted by an independent third party. It should be kept in mind that the purpose of sampling and testing for QC and acceptance is to estimate the population being produced. Depending on the definition used, the purpose of IA is to provide an independent assessment of either (1) the testing process or (2) of the product and/or the reliability of test results. Whichever definition is used, the emphasis is placed on *independent*.

QUALITY CONTROL

In early materials and methods specifications, there was often no formal QC requirement. The agency stipulated how the contractor was to perform the work and monitored the operations by inspection and testing. The testing that was done was a combination of QC and acceptance, although these terms were not generally used. As the responses indicate, that has changed somewhat under present-day materials and methods specifications in that the QC function is often performed; sometimes by the contractor and sometimes by the agency. However, irrespective of who performs the QC function, the intent is the same; that is, to assess and adjust production and construction processes so as to control the level of quality being produced in the end product. This should be a separate and distinct function from that of acceptance, which is to determine the degree of compliance with contract requirements.

However, in QA specifications that, by definition, contain ingredients of both end-result and materials and methods requirements, QC is designated as a function to be performed by the contractor. The assignment of this function to the contractor evolved from early materials and methods and statistically based specifications primarily for two reasons. The first reason was that it was found if the agency controls the contractor's process, the agency implicitly accepts responsibility for the product and must accept it, irrespective of the quality (*17,18*). The second reason was that it is the contractor's production equipment and personnel that are used to produce the material and construction and, therefore, the best entity to control these items is the contractor.

When the shift of QC responsibility from the agency to the contractor occurred, the concept and purpose of QC often became confused. This confusion apparently still exits. By definition, QC is "Those QA actions and considerations necessary to assess and adjust production and construction processes so as to control the level of quality being produced in the end product" (*13*). The key word is control, not accept. Thus, the purpose of QC is not to sample and test for acceptance. The number of questionnaire responses that indicated that agencies use the same test procedure and point of sampling for both QC and acceptance functions lend support to the conclusion that often these functions may not be separated but, simply, conducted by different parties.

An important ingredient of QC is to develop a QC plan. "Thus, it is imperative that the contractor have a functional, responsive QC Plan" (*19*). Also, it is important that the plan realize that the purpose of QC is to measure those quality characteristics (air content, density, etc.) and to inspect those activities that affect the production at a time when corrective action can be taken to prevent appreciable nonconforming material from being incorporated in the project (*17*). "For example, while 28-day concrete cylinder strength provides useful information to both the agency and the contractor, it is not a good QC quality characteristic. By the time this quality characteristic is measured, too much production has occurred to make strength results useful as a QC tool" (*12*). (Irrespective of this admonition, the responses to the questionnaire indicated that several agencies use 28-day concrete strength for QC.) Therefore, choosing the quality characteristics that best control the quality is an important step.

A QC plan can be either contractor-specific or generic. Ideally, the plan should be contractor/operation-specific. However, many agencies choose to develop a generic plan to be used by all contractors or suppliers (*20*). In any case, the contractor should develop control limits based on the production capabilities of the specific operation. For effective QC actions, the control limits should not be based on the specification limits (*12*).

Among several important ingredients in a QC plan are requirements of the use of qualified technicians and laboratories and the use of control charts. Example QC plans for HMA and both PCC structures and PCC paving are provided in Appendices B, C, and D of the *AASHTO Implementation Manual for Quality Assurance* (*19*).

Qualified Technicians

A comment regarding the use of the terms "qualified" and "certified" technicians is warranted here. Regulation 23 CFR 637 uses the term qualified personnel, as opposed to certified. One reason that "qualified" was selected is that some agencies are prohibited by law to certify technicians unless they are state employees. Also, technician certification usually implies the use of an ongoing recertification program, although technician qualification could be a one-time event. AASHTO *Standard Recommended Practice for Technician Training and Qualification Programs* (*21*) indicates that the terms "qualification" and "technician" are meant to be generic descriptions. The *AASHTO Quality Assurance Guide Specification* (*8*) uses the term "certified technicians." It is generally understood that technicians must be qualified and that one way to ensure this is to require them to have undergone some certification procedure.

If certification is required, most agencies have either an in-house certification program or participate in a regional one. Information on this subject can be found in several documents, three of which are referenced here. The National Institute for Certification in Engineering Technologies (NICET) sponsors the first two documents in the fields of highway materials (*22*) and asphalt, concrete, and soils (*23*). The third (*24*) is an outgrowth of a workshop sponsored by the National Quality Initiative. This workshop was the result of the requirement contained in 23 CFR 637 that "After June 29, 2000, all sampling and testing data to be used in the Acceptance decision or the IA [independent assurance] program shall be executed by qualified sampling and testing personnel" (*1*). Documenting how agencies implemented this requirement is one of the reasons this synthesis is needed. The FHWA issued a 1998 memorandum that provides suggestions for the requirements of qualified technicians. This memorandum is in Appendix D (*25*).

Qualified Laboratories

23 CFR 637 defines qualified laboratories as "laboratories that are capable as defined by appropriate programs established by each SHA. As a minimum, the qualification program shall include provisions for checking test equipment and the laboratory shall keep records of calibration checks" (*1*). The June 29, 2000, date referenced in this section to qualified sampling and testing personnel also applied to the use of qualified laboratories. The requirement also states that "After June 30, 1997, each SHA shall have its central laboratory accredited by the AASHTO Accreditation Program or a comparable laboratory accreditation program approved by the FHWA." And furthermore, "After June 29, 2000, any non-SHA laboratory which performs IA sampling and testing shall be accredited in the testing to be performed by the AASHTO Accreditation Program or a comparable laboratory accreditation program approved by the FHWA." This date also pertains to any non-SHA laboratory used in dispute resolution sampling and testing (*1*). The FHWA offered suggestions on the requirements of qualified laboratories in a 1998 memorandum. This is provided in Appendix D (*26*).

Statistical Process Control

One of the tools used by many manufacturing industries to help control the quality of their product is Statistical Process Control (*27*); a tool that has not been readily accepted in the highway industry. One important tool in Statistical Process Control is the use of control charts. Although many agencies require control charts to be plotted and maintained, contractors tend to comply reluctantly. This synthesis, through instruction of QA courses and interviews with practitioners, found some typical evidence of this reluctance. For instance, some of the practices used by contractors are to

- Conduct only the minimum tests required;
- Plot results on the charts at a convenient time rather than immediately, so as to react to out-of-control product;
- Use simplistic and less effective types of control charts, called "run charts";
- Not establish effective control limits;
- Use specification limits for control limits (this is sometimes an agency requirement);
- Not react when product appears to be out of control; and
- Use agency acceptance test results for their QC.

Although the use of control charts is beyond the scope of this synthesis, the purpose of their use is to provide a visual depiction of the population being produced. This means, ideally, control charts should provide estimates of the two population parameters, the average and the variability. For the reader interested in more information on control charts, including the calculation and use of control limits, the *ASTM Manual on Presentation of Data and Control Chart Analysis* can be a useful reference (*28*).

ACCEPTANCE

Because the acceptance function evolved from the earliest specifications, it is often considered the most important of the three. However, because QA is considered a system, all functions are important and should work together. The purpose of acceptance is to assess the quality of the product and, when appropriate, establish payment (*12*). This seems to be a straightforward goal but, in actuality, acceptance is a many-faceted function. Some of the considerations involved in the acceptance function are

- Acceptance procedures and requirements,
- Quality measures used,
- Possible use of contractor test results in the acceptance decision,
- Verification testing when contractor test results are used, and
- Risks to the agency and the contractor.

Acceptance Procedures and Requirements

There are many important acceptance procedure issues that must be decided on when developing the acceptance plan. As with QC, there is no single prescription that works best, but several that have been used effectively by various agencies. It is important that the agency determine what it wishes to accomplish with the acceptance plan. If the primary function is to ensure that the contractors do not totally disregard quality, then the presence of an agency inspector accompanied by a minimal amount of acceptance testing may be sufficient. This limited effort, however, will not really allow the agency to distinguish between good and poor construction and material. To do this will require additional random sampling and testing along the lines of what has traditionally been done, or greater. If the agency wants a sound statistically based plan that will enable them to determine with a low degree of risk the quality levels that the contractor is providing, then even larger sample sizes will be required (*12*).

The evolution that has occurred in QA has affected not only the relationship among the QA functions, but has taken place within a function as well. For instance, acceptance testing once concentrated on those quality characteristics that were easiest to measure; for example, gradation of the completed mix for HMA and slump for PCC. More recently, quality characteristics are preferred that affect performance; for example, volumetric properties of HMA and permeability for PCC.

Contractor Test Results Used in the Acceptance Decision

An important step in the evolution of QA programs occurred when 23 CFR 637 allowed contractor test results to be used in the acceptance decision. Skeptics often said, "You're locking the fox up in the hen house." However, recent history has indicated that, with the checks and balances required in 23 CFR 637, more testing in the acceptance function is being done using this alternative than would have been done solely by the agency under traditional acceptance testing. Studies have indicated that quality at least equal to that obtained under traditional specifications using only agency acceptance tests can be obtained with the use of contactor tests (*29*). However, a question still exists as to the validity and value of the use of contractor tests in the acceptance function. In an attempt to answer this question, NCHRP Project 10-58 (02) Using Contractor-Performed Tests in Quality Assurance is being undertaken.

Verification Testing

Verification Procedures

The ability of the comparison procedure to identify differences between two sets of test results depends on the number of tests that are being compared. The greater the number of test results in each set, the greater the ability of the procedure to identify statistically valid differences. Although a rule of thumb is a minimum agency rate of 10% of the contractor's testing rate, it is preferred to conduct a risk analysis to determine if a higher rate is warranted (*20*). It also must be decided whether it is the process or the test method that is to be verified. This relates to the use of independent or split samples.

Definitions of Hypothesis and Level of Significance

In general, a hypothesis is a statement of an assumption about a set of data. As used in this synthesis, the set of data comprises a population. The null hypothesis, H_o, defines an assumed set of conditions. (The null hypothesis can only be proven true by testing the entire population; it cannot be proven true from a sample. However, from a sample it can be shown, with specified risks, to be untrue.) The alternative hypothesis, H_α, is the hypothesis that is accepted when the null hypothesis is disproved (i.e., rejected). The level of significance, α, is the probability of rejecting a null hypothesis when it is actually true. Typical levels of significance are 0.10, 0.05, and 0.01. If for example $\alpha = 0.01$ is used and the null hypothesis is rejected, then there is only 1 chance in 100 that H_o is true and was rejected in error.

Verification of Contractor Test Results

When contractor test results are used in the acceptance decision, the preamble of the final rule 23 CFR 637 requires, in addition to the IA program, a verification program including the use of independent samples for the verification sampling and testing, and specifically states that

> There are three sources of differences between two test results, differences in the material, differences in test procedures, and differences in sampling procedures. Split samples will only address the differences in test procedures and will only provide assurance that the contractor is performing the test properly. In a balanced system it is also necessary to assure that sampling of materials is performed properly. It is our intent that the verification sampling and testing program be used to independently validate the quality of the material. Using independent samples will insure that all sources of differences are measured. The FHWA recognizes the need to ensure that each contractor performs the tests correctly; that is the reason for extending laboratory and testing personnel qualification requirements and IA program requirements to the contractor if the contractor's test results are to be used in the acceptance decision. The FHWA expects the testing variability between the contractor and the State to be held to a minimum by requiring the contractor's program to be covered by an IA program and requiring the testing personnel and laboratories to be qualified. The FHWA has changed the definition of "verification sampling and testing" and Sec. 637.207(a)(1)(ii)(B) to clarify the fact that the verification sampling and testing program is being used to validate the quality of the material (*1*).

Even with the above explanation provided in 23 CFR 637, there exists misunderstanding about the difference in information provided by the use of independent versus split samples. In an effort to clarify the difference, the manual *Optimal Procedures for Quality Assurance Specifications* uses the terms "Test Method Verification" for the analysis using split samples and "Process Verification" for the analysis using independent samples (*12*).

Process Verification Procedures

There are two procedures that appear in the OPQAS and the *AASHTO Implementation Manual for Quality Assurance* for verification of independently obtained test results (*12,19*). The tests most often used are the *F*-test and *t*-test, which are usually used together. However, a procedure that compares a single agency test result with 5 to 10 contractor test results is also used. Both of these are discussed here (*12*).

F-test and t-test This procedure involves two hypothesis tests, where the null hypothesis, H_o, for each test is that the contractor's tests and the agency's tests are from the same population. In other words, the null hypotheses are (1) that the variabilities of the two data sets are equal for the *F*-test, and (2) that the means of the two data sets are equal for the *t*-test.

It is important to compare both the means and the variances when comparing two data sets. A different test is used for each of these comparisons. The *F*-test provides a method for comparing the variances (standard deviations squared) of the two sets of data. Differences in means are assessed by the *t*-test.

The procedures involved with the *F*-test and *t*-test may at first seem complex. The *F*-test and *t*-test approach also requires more than one agency test result before a comparison can be made. These reasons may persuade an agency to seek a simpler approach. However, the combination of the *F*-test and *t*-test is much more statistically sound and has more power to detect actual differences than the second method that relies on a single agency test for the comparison. Any comparison method that is based on a single test result will not be as effective in detecting differences between data sets.

Some of the complexity of the *F*-test and *t*-test comparisons can be eliminated by the use of computer programs. Many spreadsheet programs can conduct these tests.

Single Agency Test Results Compared with a Number of Contractor Test Results In this method, a single agency test result is compared with 5 to 10 contractor test results. The single agency test result must fall within an interval that is defined from the mean and range of the 5 to 10 contractor test results. The allowable interval within which the agency test result must fall is

$$\overline{X} \pm CR$$

where $\overline{X}$ and R are the mean and range, respectively, of the contractor test results, and C is a factor that varies with the number of contractor test results (*12*).

In discussing this procedure, the OPQAS recommends that

> this method should not be used. This method was developed to be very simple. It suffers from the fact that only a single agency test is used when making the comparison. Any method that relies on a single data value will not be very powerful at detecting differences. This is due to the high variability that is associated with individual, as compared with mean, values.
>
> For example, if the standard deviation for measuring air content in PCC is 0.75 percent, then for a comparison based on five contractor tests, there is only about a 33 percent chance of detecting an actual difference of 2.25 percent [3 standard deviations] between contractor and agency means. The chance only increases to about 57 percent when 10 contractor tests are used (*12*).

Although this is not a particularly efficient approach and is not recommended, the responses to the questionnaire indicate that it is being used by several agencies. This is an indication that many agencies favor simpler approaches. The tendency to use simple measures as opposed to more effective ones is not new. Afferton et al. (*30*) discussed this issue emphatically in 1992:

> In the field of quality assurance, the problem manifests itself in a particularly troublesome way. In a discipline dedicated to the pursuit of excellence, it seems totally inappropriate to tolerate specifications and consensus standards that are far from excellent. If demands for excellence are to be made of the construction industry, it is imperative that engineers be willing to demand the same of themselves in the development of the specifications and standards that govern the work.
>
> Ironically, all the necessary statistical tools are well developed and readily available. Those unfamiliar with the mathematical principles underlying Statistical Quality Analysis (SQA) procedures may find it difficult to realize just how inadequate many current practices are.
>
> Leaders within the transportation field must invite an open and thorough scrutiny of current practices and must insist that improvements be made where necessary. To do anything less would be a breach of both professional ethics and public trust.

Test Method Verification Procedures

These procedures involve the comparison of split sample values. And, just as for process verification, there are two procedures used most often for comparing split samples. These are the D2S limits and the paired *t*-test. Comparing split samples is called test method verification (23 CFR 637 uses the term comparison) to separate it from the comparison of independent samples, called process verification, and discussed previously under Verification Testing (*12*).

D2S Limits This is the simplest procedure that can be used for verification, although it is the least powerful. Because the procedure uses only two test results, it cannot detect real differences unless the results are far apart. The value provided by this procedure is contained in many AASHTO and ASTM test procedures. The D2S limit indicates the maximum acceptable difference between two results obtained on test portions of the same material (and thus applies to only split samples), and is provided for single and multi-laboratory situations. It represents the difference between two individual test results that has an approximately 5% chance of being exceeded if the tests are actually from the same population.

When this procedure is used for test method verification, a sample is split into two portions and the contractor tests one split-sample portion, although the agency tests the other split-sample portion. The difference between the contractor and agency test results is then compared with the D2S limits. If the test difference is less than or equal to the D2S limit, the contractor test result is considered verified (compared). If the test difference exceeds the D2S limit, then the contractor's test result is not verified (compared) and the source of the difference is investigated.

Paired t*-test* For the case in which it is desirable to compare more than one pair of split-sample test results, the *t*-test for paired measurements can be used. This test uses the differences between pairs of tests and determines whether the average difference is statistically different from 0. Therefore, it is the difference *within* pairs, not between pairs, that is being tested. The *t*-statistic for the *t*-test for paired measurements is

$$t = \frac{\left|\overline{X}_d\right|}{\frac{s_d}{\sqrt{n}}} \qquad (1)$$

where

$\overline{X}_d$ = average of the differences between the split-sample test results,

s_d = standard deviation of the differences between the split-sample test results, and
n = number of split-sample differences.

The calculated t-value is then compared with the critical value, t_{crit}, obtained from a table of t-values at a level of $\alpha/2$ and with $n - 1$ degrees of freedom. Computer spreadsheet programs contain statistical test procedures for the paired t-test. This makes the implementation process straightforward.

Risks and Operating Characteristic Curves

Establishing the limits to be used for acceptance is an important part of a QA program. Making the limits too restrictive deprives the contractor of a reasonable opportunity to meet the specification. Making them not sufficiently restrictive makes them ineffective in controlling quality. Selection of the limits relates to the determination of risks. The two types of risk encountered are the seller's (or contractor's) risk, α, and the buyer's (or agency's) risk, β. A well-written QA acceptance plan takes these risks into consideration in a manner that is fair to both the contractor and the agency. Too large a risk for either party undermines credibility. Therefore, the risks should be both reasonably balanced and reasonably small.

The two types of risk, α and β, are very narrowly defined to occur at only two specific quality levels. β is the probability of accepting material that is exactly at the rejectable quality level, whereas α is the probability of rejecting material that is exactly at the acceptable quality level. These definitions do not, therefore, provide an indication of the risks over a wide range of possible quality levels. To evaluate how the acceptance plan will actually perform in practice, it is necessary to construct an operating characteristic (OC) curve that is a graphic representation of an acceptance plan that shows the relationship between the actual quality of a lot and either (1) the probability of its acceptance (for accept/reject acceptance plans) or (2) the probability of its acceptance at various pay levels (for acceptance plans that include pay adjustment provisions) (*13*).

Although the subjects of risks and OC curves are very important aspects of QA programs, the survey questionnaire did not cover the establishment or use of either. Therefore, for more information on these subjects, a discussion of risks and OC curves is provided in Appendix E. For additional information on risks and OC curves, refer to the OPQAS manual (*12*).

Quality Measures

Before listing the many quality measures used by agencies, a brief discussion of three of the most often used, individual values, average, and percent within limits (PWL) [or percent defective (PD)] is warranted. Although there are several quality measures that are used for acceptance, some are more effective than others in providing an estimate of the population. The earliest sampling and testing results relied on decisions based on individual values. There was no accumulation of values to determine an average, nor a measure of variability. The results of the AASHO Road Test showed that this simple method of examining test results was inadequate. In the 1960s, with the initiation of statistically based specifications, acceptance specifications termed "Variability Known" were popular. These specifications measured only the average and assumed the variability to be known or constant. Several agencies quickly learned that in the highway industry this is an inaccurate assumption, because variability is rarely known or constant. As will be seen from the survey results, some agencies still use only the average as the preferred quality measure.

Some of the great advances in QA specifications were made in the 1970s. One was an understanding and analysis of risks, both to the contractor (seller) and to the agency (buyer). The advantages in the use of statistically based specifications had been recognized, logically, since their early development. However, with the introduction of the concept of risks, the advantages of a well-written statistically oriented specification to both the seller and buyer in terms of balanced risks became quantifiable (*7*).

Another advancement made in this decade was the adoption by some state agencies, such as New Jersey and Pennsylvania, of "Variability Unknown Specifications." In this type of specification, the average and variability are combined to estimate a quality level. Acceptance plans for this type of specification are based on procedures found in Military Standard 414 (*31*). The application of the specification is in the form of PWL or PD.

Typical quality measures used, in addition to the average, are

- Individual values—the earliest form of acceptance. Because of the large variability associated with single values, this is one of the least-effective acceptance procedures.
- Range—the difference between the largest and smallest values in a set of data. It is the simplest measure of variability and is a reasonably effective measure for small sample sizes (*28*).
- Standard deviation—a measure of the dispersion of a series of results around their average. The standard deviation is the typical measure of variability and is a measure of precision.
- PWL—the percentage of the lot falling above the lower specification limits (LSL), beneath the upper specification limits (USL), or between the USL and LSL.
- PD—the percentage of the lot falling outside specification limits.
- Average absolute deviation (AAD)—for a series of test results, the mean of absolute deviations from a target or specified value. Because this quality measure uses a target value as a reference point, it is not usually applied

with a quality characteristic with a single specification limit; for example, concrete compressive strength.
- Conformal index (CI)—a measure of the dispersion of a series of results around a target or specified value; this is a measure of accuracy. This quality measure is not usually applied with a quality characteristic with a single specification limit, for the same reason that AAD is not.

Training and Certification

Just as training and/or certification are necessary for QC purposes, they are also necessary for acceptance sampling and testing. Regulation 23 CFR 637, and the 1998 letter from the FHWA suggesting ways to implement 23 CFR 637, "requires that all sampling and testing of highway materials for Federal-aid projects on the National Highway System (NHS), subsequent to June 29, 2000, must be performed by qualified technicians."

INDEPENDENT ASSURANCE

Because at least two different definitions of IA are used, there has been some confusion as

- To the purpose of IA,
- How the IA program should be conducted, and
- What the comparison of test results reveals.

Depending on which definition is used, the purpose of IA is to provide an independent assessment of the test results obtained from QC and acceptance or, in the broader context, to provide an independent assessment of the product and/or the reliability of test results obtained from the process control and acceptance testing. The survey results indicate that some agencies use one definition and some the other. In both cases, the intended purpose of IA is to provide a connection to the acceptance plan. It involves a separate and distinct schedule of sampling, testing, and observation. The survey results also indicate that in many agencies IA personnel perform other functions in addition to those related to IA. When statistical comparisons are made, they can provide an assessment of split-sample test results. The results from these comparisons are intended to reveal whether or not the test results from either QC or acceptance are statistically comparable to the independent test results.

Several agencies have conducted studies to determine the effectiveness of the IA program (*32–34*). The results typically have suggested ways the agencies have improved and can continue to improve their IA systems.

It is important that an IA program compare results and detect deficiencies, when they exist, in a timely manner. This improves the reliability of testing results. The timely comparison of data may be restricted by agency resources, including personnel, facilities, and geographic constraints. These resource needs must be considered in the IA program. The importance of the use of qualified personnel to conduct the IA tests has been mentioned previously. Important issues that are addressed in the survey include:

- How the IA unit in each agency is organized,
- How many full-time equivalent (FTE) personnel are used in the IA function, and
- To what testing the agency applies the IA function.

The previous discussion provides an indication of the complexity of the IA function. Attempting to cover its many aspects in a questionnaire and in this synthesis was a challenging task.

USE OF CONSULTANTS AND INNOVATIVE PRACTICES IN QUALITY ASSURANCE PROGRAMS

Many agencies continue to downsize, restructure their organizations, and, as a consequence, reduce personnel levels (*35*). To address these issues, agencies have taken several steps to relieve the pressure on their remaining personnel. Two such steps are the use of consultants for testing and inspection and the use of innovative practices. One of the innovative practices is the use of warranties (*35–37*). Also, design–build is being tried by some agencies (*29,38*). Although these are different forms of contracting than the typical materials and methods or QA types, they still involve the control and acceptance of the materials and construction. This synthesis sought to determine how agencies that use these newer procedures conduct QC and acceptance functions. Specifically, does the agency require a different procedure than if the typical specification was used and are the QC and acceptance procedures required at all?

Chapter nine discusses the agencies' responses to the use of QA procedures when consultants or innovative practices are used. The responses to this section were typically short and did not contain much detailed discussion.

PAY ADJUSTMENT SYSTEMS

An advancement in the 1970s was the adoption of the concept of incentive pay clauses for product that was exceptionally better than required by the specifications. This concept was complementary to the concept of disincentive pay clauses previously used. Benefits of incentive pay clauses were viewed as improved quality, the positive psychological effect of being rewarded for excellent control, and fairness to the contractor (*4*).

One of the primary purposes of a payment schedule is to provide payment commensurate with the quality provided. Often this includes sufficient incentive to produce the desired level of quality at the time of initial construction. Effective payment schedules encourage contractors to apply appropriate QC measures to ensure that the finished product will equal or exceed the desired level of quality a high percentage of the time. The rationale of the agency is that the small additional cost of good QC practices expended in advance is better than being faced with the anticipated future costs of poor quality construction, which may lead to premature failure of pavements, excessive maintenance repairs, possibly unsafe driving conditions, etc. (*12*).

A secondary purpose of the payment schedule is to recoup at least part of the anticipated future costs that are likely to occur when poor quality is received. For a variety of reasons, there will occasionally be times when QC measures are either absent or ineffective, leading to less than satisfactory work. Provided the work is not too seriously deficient, it usually is both impractical and unnecessary to require removal and replacement (accept/reject), and the better solution in these cases is to accept the work at a reduced price. This is consistent with the legal principle of liquidated damages, a well-established means for recovering losses that are difficult to quantify precisely at the time the contract is executed (*12*).

As the questionnaire confirmed, there are several types of acceptance procedures being used, including pay adjustment schedules and the older accept/reject procedure. The accept/reject procedure is still used extensively for an entire material/construction item, such as soils and embankments. However, it is also used extensively as "screening tests" for a material as it is incorporated in the construction; for example, air content for PCC and temperature for HMA.

For pay adjustment schedules either step pay factors or equations are typically used. The earliest payment schedules were usually stepped schedules, such as that shown in Table 1 and plotted in Figure 1 (*12*).

More recently, there has been a tendency to use continuous (equation-type) payment schedules. One is shown in Eq. 2 and also plotted in Figure 1.

$$PF = 55 + 0.5\,PWL \qquad (2)$$

where

PF = payment as a percent of the unit bid price, and
PWL = estimated percent within limits.

TABLE 1
TYPICAL STEPPED PAYMENT SCHEDULE BASED ON PWL (*12*)

Estimated PWL	Payment Factor, %
95.0–100.0	102
85.0–94.9	100
50.0–84.9	90
0.0–49.9	70

Note: PWL = percent within limits.

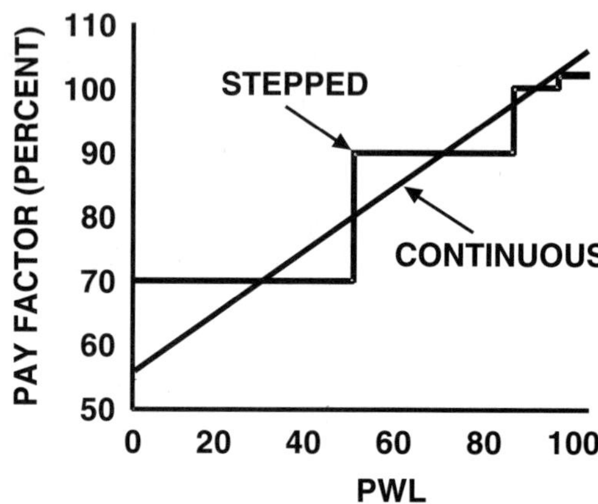

FIGURE 1 Example of stepped and continuous payment schedules (*12*). PWL = percent within limits.

Although risk analysis could show these two payment schedules to have very nearly the same long-term performance, especially for small sample sizes, there is a distinct advantage associated with the continuous form. When the true quality level of the work happens to lie close to a boundary in a stepped payment schedule, the quality estimate obtained from the sample may fall on either side of the boundary owing primarily to chance. Depending on which side of the boundary the estimate falls, there may be a substantial difference in payment level, which may lead to disputes over measurement precision, round-off rules, and so forth. This potential problem can be avoided with continuous payment schedules that provide a smooth progression of payment as the quality measure varies (*12*).

CHAPTER THREE

QUALITY ASSURANCE PROGRAMS FOR SOILS AND EMBANKMENTS

QA programs used to control and accept soils and embankments tend to differ from those of other materials. The responses from the questionnaire confirm that the QA programs for soils and embankments are not as rigorous as for other construction products. This is primarily the result of the variability of these materials. As defined, statistically based specifications are those in which properties of the desired product or construction are described by appropriate statistical parameters. Soils and embankments tend not to meet this requirement because of their large degree of heterogeneity. It may be argued that a testing program that included a large number of tests could estimate the larger degree of heterogeneity. However, because more testing is not done, it is assumed that agencies do not see this option as being cost-effective. As McMahon et al. indicate in "Quality Assurance in Highway Construction" (*39*), the variability of the material itself impedes the use of overall standard deviation as a measure of contractor performance. As the composition of the material becomes more variable, results of the compaction process also become more variable. Data from a California report indicate how the variability can differ from soil to soil. Standard deviations of relative compaction on three projects were 2.44%, 3.09%, and 5.52% for a relatively homogeneous fine-grained soil, a soil with intermediate variable properties, and a very heterogeneous soil, respectively (*40*). A study of embankment compaction and moisture content undertaken by the Minnesota Department of Highways found that:

> The large standard deviations of the data obtained indicate the wide variation or dispersion for the characteristics measured. Part of this variability can be attributed to test methods. However, the entire variability cannot be attributed to testing error as there may be differences in the material placed or densities when the tests were taken. In any case, the variation in density and moisture content of embankments has been found to be much greater than had been expected when this phase of the research program was initiated. Much of the variation may be in the contractor's process. When informed that random sampling would be used this should have had some psychological effect for the better construction of the embankment. With this in mind, there is a possibility that even more variation exists than what was determined in this portion of the study (*41*).

TYPE OF QUALITY ASSURANCE PROGRAM

QA programs other than statistically based ones predominate for soils and embankments. This was borne out by the questionnaire responses. As Figure 2 shows, of the 45 agencies responding, 25 use primarily materials and methods provisions in their QA programs, 23 use QA programs with the agency controlling quality and performing acceptance, 16 use QA programs with the contractor controlling quality and the agency performing acceptance, and only 6 (and one pilot testing) use QA programs with the contractor controlling both the quality and contractor tests used in the acceptance decision.

Of the 45 respondents, 21 require the same test methods for QC and acceptance and 13 use the same point of sampling. Only one agency requires a different test method for QC and acceptance and 23 agencies specify tests only for acceptance. Fifteen agencies use different points for sampling for QC and acceptance; four use independent random locations for both, two let the contractor choose the point for QC sampling, and two perform QC tests at the source and acceptance at the road.

QUALITY CONTROL

Figure 3 shows that the attributes used most often for QC of soils and embankments are moisture content and compaction. Of the 45 respondents, 17 use moisture content and 18 use compaction.

For the frequency of QC tests, 19 agencies require an agency-established frequency and 5 require that the contractor choose the frequency. Additional evidence that these materials tend to be heterogeneous is that no agency requires control charts, although three have QC requirements for gradation.

ACCEPTANCE

The attributes used most often for acceptance of soils and embankments are also moisture content and compaction. Figure 3 shows that 44 of the 45 respondents accept soils and embankments based on compaction and 29 based on moisture content. Some of the lesser-used acceptance attributes are gradation, Atterberg limits, AASHTO classification, maximum lab density, proof rolling, volume change, *R*-value, and depth of lift.

All 45 agencies reported that they use an accept/reject acceptance plan. Of these, six use contractor test results as part of the acceptance decision. The verification system used

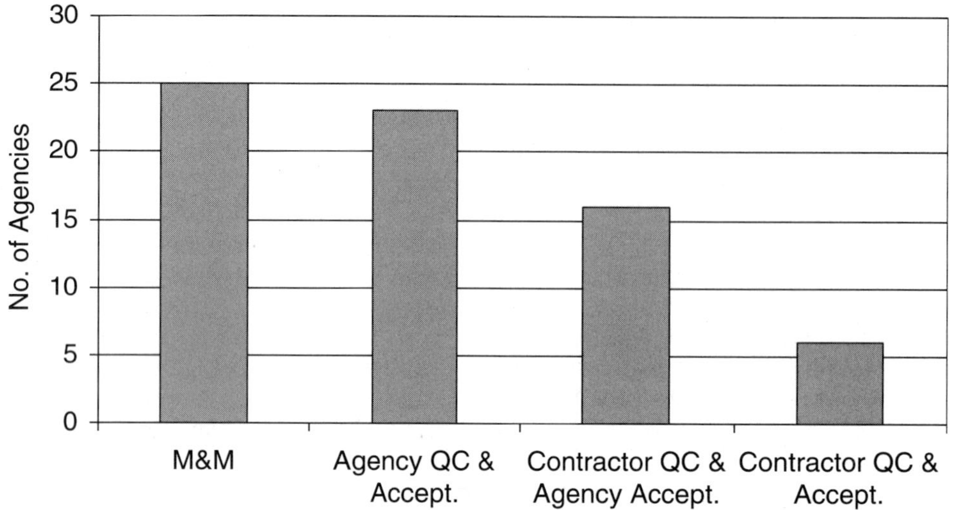

FIGURE 2 Types of QA programs used for soils and embankments (45 responses). M&M = materials and methods.

is based either on one contractor test to one verification test, used by four agencies, whereas three use agency-established tolerances, and one AASHTO D2S tolerances. Five agencies use one agency test compared with several contractor tests. Three agencies base the comparison on a lot and one bases it on a complete project. Three use independent samples, two use split samples, and one uses both. (These numbers indicate that some agencies use more than one procedure. Although the questionnaire was not "attribute specific," it is likely that different procedures are used for different attributes.)

QUALITY MEASURES USED FOR ACCEPTANCE

Figure 4 shows that the most common quality measure for soils and embankments, used by 34 agencies, is an individual value. This reflects McMahon et al.'s experience in the FHWA report (*39*). However, eight agencies use the range, three use the average, and three use PWL.

TRAINING AND CERTIFICATION

Training and certification for agency personnel involved in inspection, sampling, and testing of soils and embankments are done primarily in-house. Figure 5 shows that 32 agencies require in-house training and 22 require in-house certification. Seven agencies use regional programs for agency personnel training and 11 use regional certification. Other training programs used are NICET and college or university training. For contractor personnel, most likely because of the large number of agency-oriented QA programs, only 12 agencies require in-house training and certification for contractor personnel. Additionally, five use regional certification, two use regional

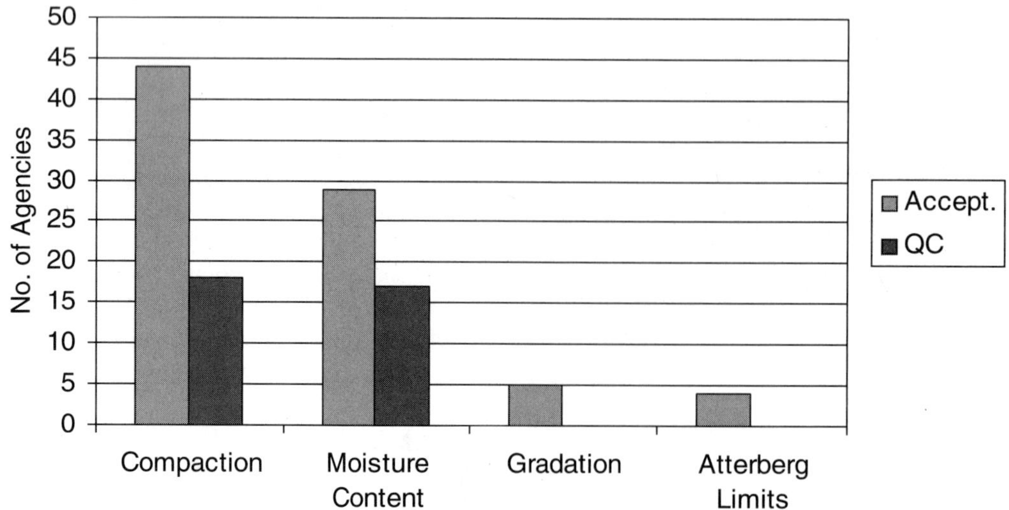

FIGURE 3 Attributes used for QC and acceptance of soils and embankments (45 responses).

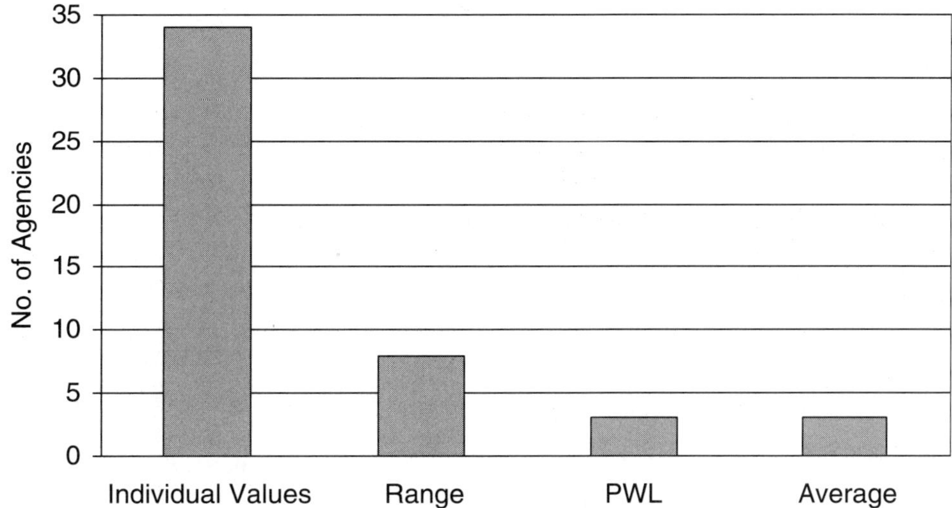

FIGURE 4 Quality measures used for acceptance for soils and embankments (45 responses). PWL = percent within limits.

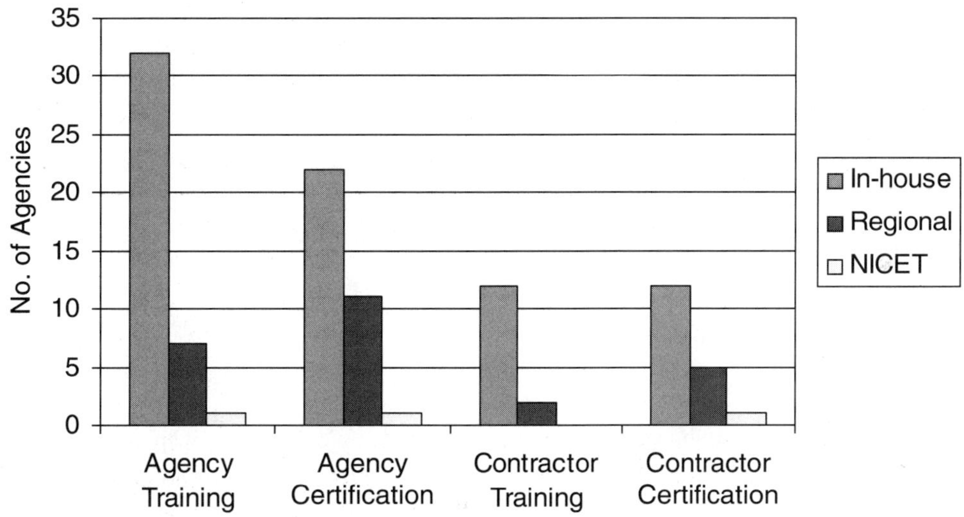

FIGURE 5 Training and certification requirements for soils and embankments (44 responses). NICET = National Institute for Certification in Engineering Technologies.

training for contractor personnel, and three use university or joint agency–industry certification.

CANADIAN QUALITY ASSURANCE PRACTICES

The Canadian QA practices for soils and embankments are similar to those used in the United States. For types of QA programs they use primarily materials and methods provisions. The QC and acceptance attributes used are generally based on moisture content and compaction and, similar to the U.S practices, they tend to use the same test methods and point of sampling for both QC and acceptance. Four provinces use an agency-established frequency for QC tests. Two provinces use PWL, two use individual values, and one uses individual values and averages for acceptance. Of the five provinces responding, all use accept/reject/rework provisions for acceptance and none use contractor tests in the acceptance decision. Training is required only for agency personnel and in-house programs are used.

CHAPTER FOUR

QUALITY ASSURANCE PROGRAMS FOR AGGREGATE BASE AND SUBBASE

QA programs used to control and accept aggregate base and subbase tend to be similar to those for other processed materials (*42*). Indeed, some agencies require these materials to be processed through a plant similar to HMA (*43*). However, other agencies do not have this requirement. This is one reason some agencies use statistically based specifications and others tend to use materials and methods provisions in their QA programs.

TYPE OF QUALITY ASSURANCE PROGRAM

Of the 45 respondents, 15 use materials and methods provisions, 14 use QA programs with the agency controlling quality and performing acceptance, 21 use QA programs with the contractor controlling the quality and the agency performing the acceptance, and 10 use QA programs with the contractor controlling the quality and the agency using contractor test results in the acceptance decision (Figure 6).

Figure 7 shows that 29 agencies require the same test methods for QC and acceptance, 3 require a different test method for QC and acceptance, and 19 specify test methods only for acceptance. Fifteen agencies use the same point of sampling for QC and acceptance, with 18 using different points for sampling. Of these, four use random locations for both, eight have QC samples taken at the crusher, five sample for acceptance from the road, and two let the contractor choose the point for QC sampling.

QUALITY CONTROL

Table 2 and Figure 8 show the attributes used for both QC and acceptance of aggregate base and subbase.

The attributes most often used for QC are gradation (27 agencies), compaction (20), and moisture content (14). One lesser-used QC attribute is aggregate fractured faces, which is used by nine agencies.

For the frequency of QC tests, 27 agencies use an agency-established frequency and 8 let the contractor choose the frequency. In addition, agencies require the use of control charts.

ACCEPTANCE

Table 2 and Figure 8 show that the attributes used most often for acceptance of aggregate base and subbase are compaction, which is used by all 45 agencies, and gradation used by 42. Twenty-four agencies accept aggregate base and subbase based on moisture content, and 21 accept on aggregate frac-

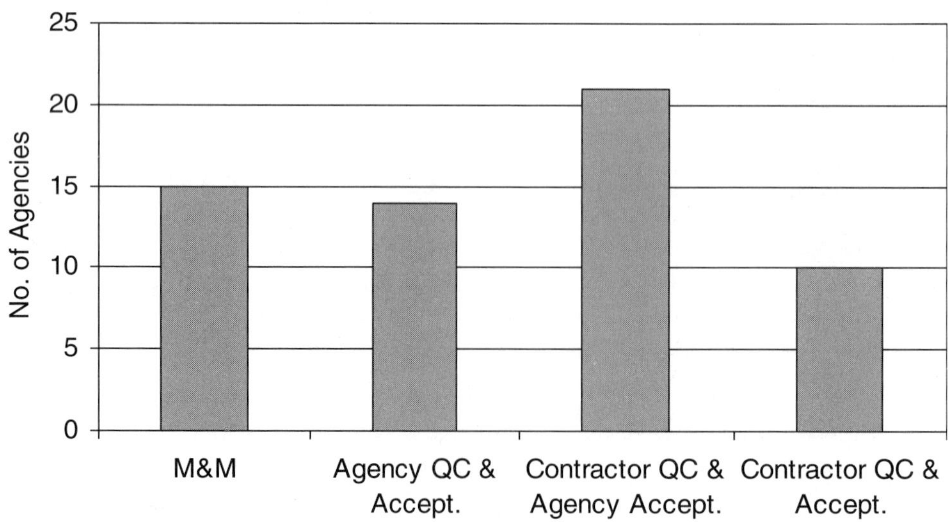

FIGURE 6 Types of QA programs used for aggregate base and subbase (45 responses). M&M = materials and methods.

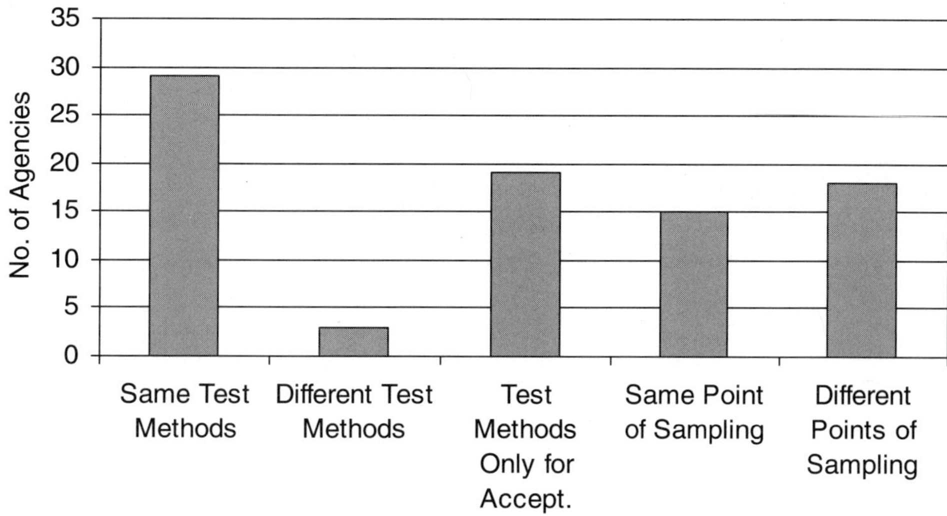

FIGURE 7 Test methods and point of sampling used for aggregate base and subbase (45 responses).

tured faces. Some of the lesser-used acceptance attributes are Atterberg limits, LA abrasion, thickness, sand equivalence, and *R*-value.

Thirty-five agencies use an accept/reject acceptance plan, whereas 16 use a pay adjustment system. Of the 16 that use pay adjustment, 8 use a stepped pay schedule, 4 use equations, and 3 use other procedures. No single agency uses only an incentive, 11 use only a disincentive, and 5 use both. Thirteen agencies use contractor test results as part of the acceptance decision. For procedures where the contractor test results are used in the acceptance decision, five agencies use all attributes, three use attributes based on only accept/reject, one uses only attributes that do not involve pay, and one uses pay reduction for gradation.

The verification system used is either based on one contractor test to one verification test, used by six agencies, AASHTO D2S (two), agency-established tolerances (one), and other verification procedures (three). Five agencies use *F*- and *t*-tests for a comparison with accumulated tests. Three agencies use one agency test compared with several contractor tests. Six agencies use the comparison based on a lot, and three use a completed project. Five agencies use independent samples, two use split samples, and three use both.

QUALITY MEASURES USED FOR ACCEPTANCE

Table 3 and Figure 9 show the quality measures used for acceptance of aggregate base and subbase. The most com-

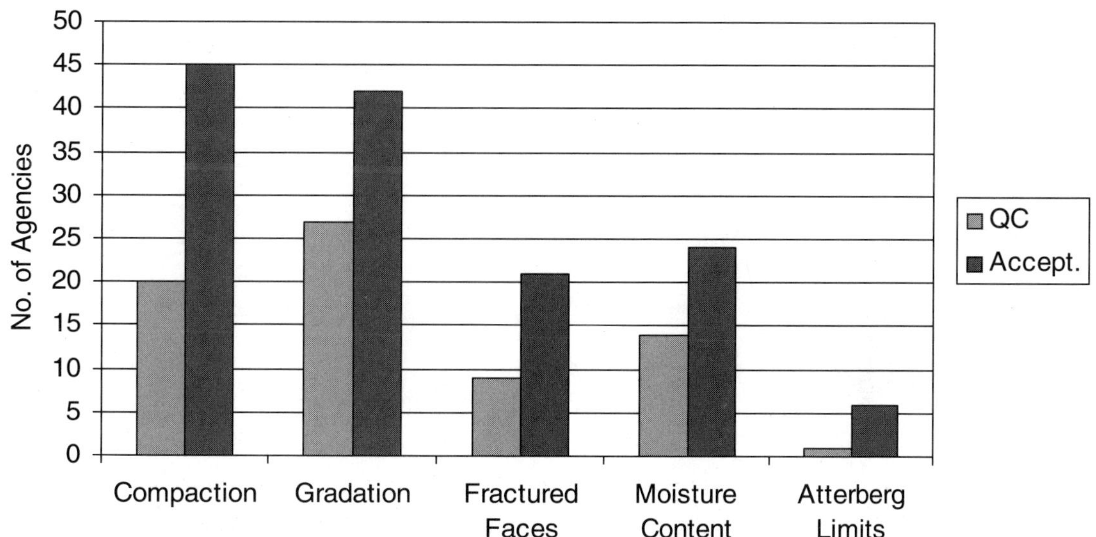

FIGURE 8 Attributes most often used for QC and acceptance of aggregate base and subbase (45 responses).

TABLE 2
ATTRIBUTES USED FOR QC AND ACCEPTANCE OF AGGREGATE BASE AND SUBBASE

Attribute	QC	Acceptance
Gradation	27	42
Aggregate fractured faces	9	21
Moisture content	14	24
Compaction	20	45
Atterberg limits	1	6
LA abrasion	0	4
Thickness	0	4
Sand equivalence	0	3
R-value	0	3

Note: 45 responses.

TABLE 3
QUALITY MEASURES USED FOR ACCEPTANCE OF AGGREGATE BASE AND SUBBASE

Quality Measure	No. of Agencies
Individual values	25
Percent within limits	13
Range	13
Average	7
Standard deviation	3
Average absolute deviation	1
Conformal index	1

Note: 45 responses.

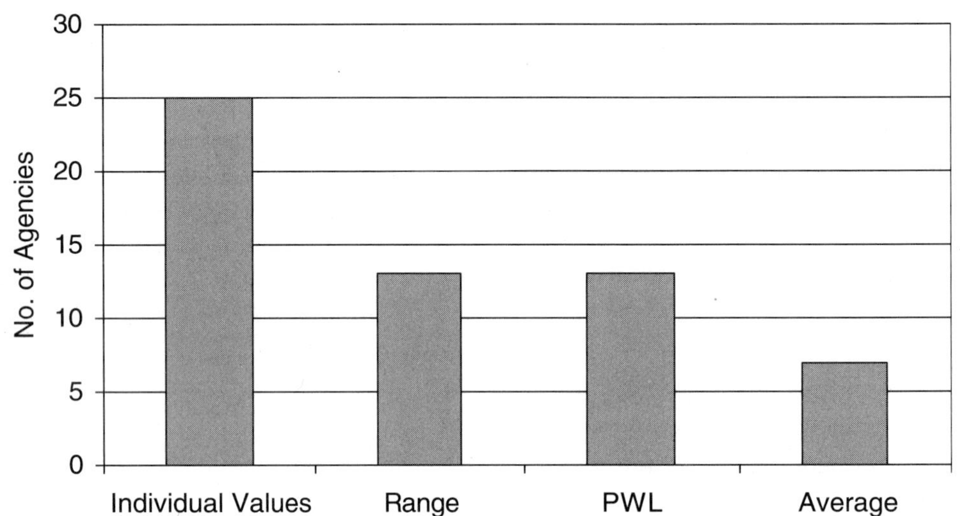

FIGURE 9 Quality measures used for acceptance of aggregate base and subbase (45 responses). PWL = percent within limits.

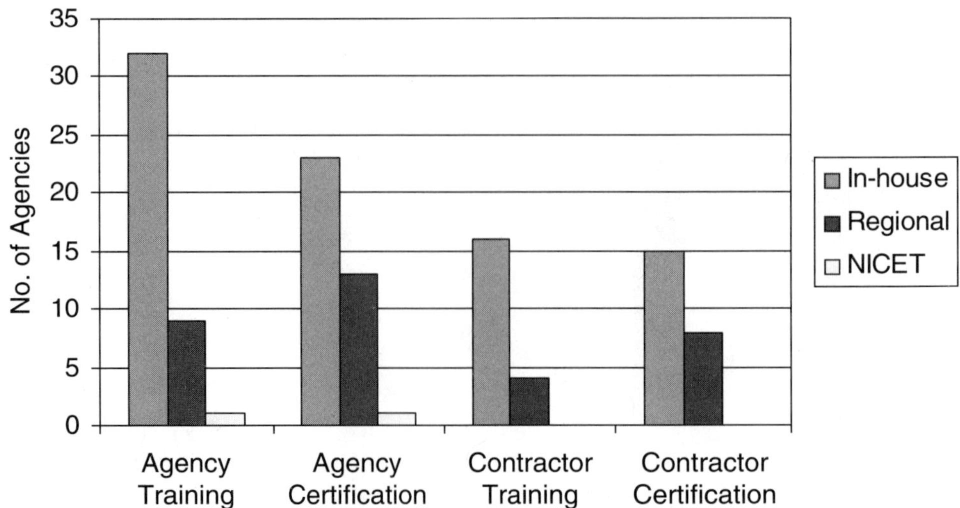

FIGURE 10 Training and certification requirements for aggregate base and subbase (45 responses). NICET = National Institute for Certification in Engineering Technologies.

TABLE 4
TRAINING AND CERTIFICATION PRACTICES USED FOR AGGREGATE BASE AND SUBBASE

	Agency Personnel		Contractor Personnel	
Program	Training	Certification	Training	Certification
In-house	32	23	16	15
Regional	9	13	4	8
NICET	1	1	0	0
Joint agency–industry	2	2	2	2
University	2	2	2	2
Crushed stone assoc.	1	1	1	1
State board	0	1	0	1

Note: 45 responses.

mon quality measure for aggregate base and subbase, used by 25 agencies, is an individual value. Thirteen agencies use PWL, seven use average, and 13 use the range, either by itself or in combination with the average. The total number of quality measures exceeds the number of responses, indicating that some agencies use more than one quality measure. For example, one agency uses PWL for cement-treated base and uses range for unbound aggregate base. Also to be noted is that the use of the conformal index was reported by one agency, West Virginia, for annual source approval of aggregates for base and subbase.

TRAINING AND CERTIFICATION

Training and certification for agency personnel involved in inspection, sampling, and testing aggregate base and subbase is very similar to that done for soils and embankments. Table 4 and Figure 10 show that 32 agencies have in-house training and 23 require in-house certification for agency personnel. Nine agencies use regional programs for agency personnel training and 13 use regional certification. One agency uses NICET for training and certification and four use university training, joint agency–industry, or the local crushed stone association for certification. For contractor personnel, 16 agencies require in-house training and 15 require in-house certification. Additionally, four agencies use regional training for contractor personnel, eight use regional certification, and four use university, the crushed stone association, state board of registration, or joint agency–industry certification.

CANADIAN QUALITY ASSURANCE PRACTICES

The Canadian QA practices for aggregate base and subbase are also similar to those used in the United States. For types of QA programs they use either materials and methods provisions or QA programs, with the contractor controlling quality and the agency performing acceptance. The QC and acceptance attributes used are generally based on gradation, aggregate fractured faces, and compaction. To a lesser extent, they use moisture content and LA abrasion. They differ from U.S. practice in that three provinces use the Micro–Deval test as an aggregate quality test. However, similar to the U.S practices, they tend to use the same test methods and point of sampling for both QC and acceptance. For QC, four provinces use an agency-established frequency for QC tests. Two provinces use PWL, two use individual values, and one uses averages for acceptance. Of the five provinces responding, four use accept/reject/rework provisions for acceptance and one uses pay adjustments. One province uses contractor tests in the acceptance decision based on accept/reject attributes. This province uses agency-established limits for verification and takes both independent and split samples. Agency personnel training is required by three provinces using in-house programs and one agency requires regional certification for contractor personnel.

CHAPTER FIVE

QUALITY ASSURANCE PROGRAMS FOR HOT-MIX ASPHALT

QA programs used to control and accept HMA tend to differ from those for other processed materials in the degree of usage of statistically based specifications. This type of specification and the associated QA programs have been under development and used longer and by more agencies than those for other materials (*44–52*). Therefore, as confirmed by the questionnaire responses, materials and methods specifications are seldom used for HMA.

TYPE OF QUALITY ASSURANCE PROGRAM

As Figure 11 shows, only two agencies still use materials and methods provisions for HMA. Twenty-one agencies use QA programs with the contractor controlling quality and agency performing acceptance, and 25 use QA programs with the contractor controlling the quality and the agency using contractor test results in the acceptance decision. This is the largest number of agencies using this procedure for any material.

Figure 12 shows that 41 agencies require the same test methods for QC and acceptance and 24 use the same point of sampling. One agency requires a different test method for QC and acceptance and 10 agencies specify test methods only for acceptance. Seventeen agencies use different points for sampling for QC and acceptance, seven use independent random locations for both, two let the contractor choose the point for QC sampling; and four have QC samples taken at the plant and sample for acceptance from the road.

QUALITY CONTROL

Table 5 and Figure 13 show the attributes used for both QC and acceptance of HMA. The attributes used most often for QC are asphalt content, gradation, and compaction. Forty agencies use asphalt content, 43 use gradation, and 28 use compaction. Other often used QC attributes are volumetric properties, ride quality, aggregate fractured faces, and thickness. Lesser-used attributes for QC are sand equivalence, aggregate percent moisture, and moisture sensitivity.

For the frequency of QC tests, 34 agencies use an agency-established frequency and 12 let the contractor choose the frequency. Twenty-seven agencies require the use of control charts.

ACCEPTANCE

Table 5 and Figure 13 show that the attributes used most often for acceptance of HMA are compaction, used by 44 agencies; asphalt content, used by 40; and ride quality, used by 39.

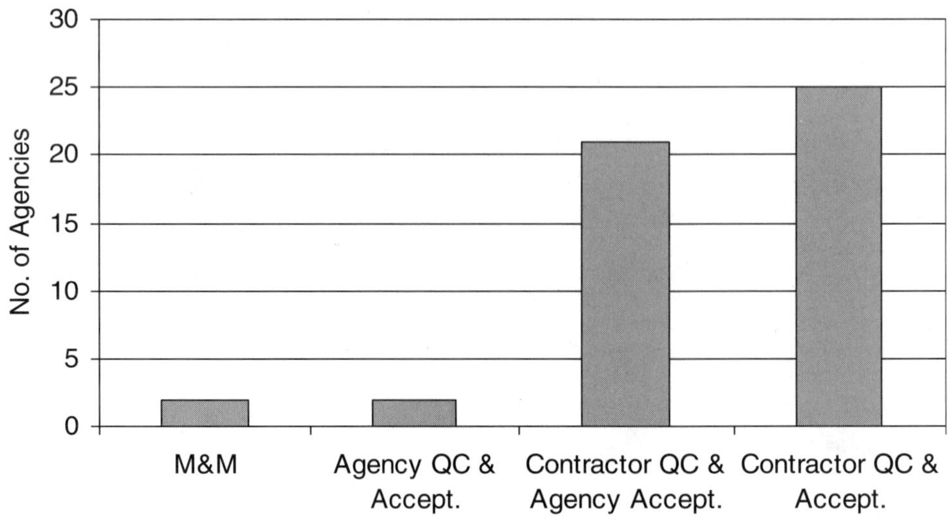

FIGURE 11 QA programs for HMA (45 responses). M&M = materials and methods.

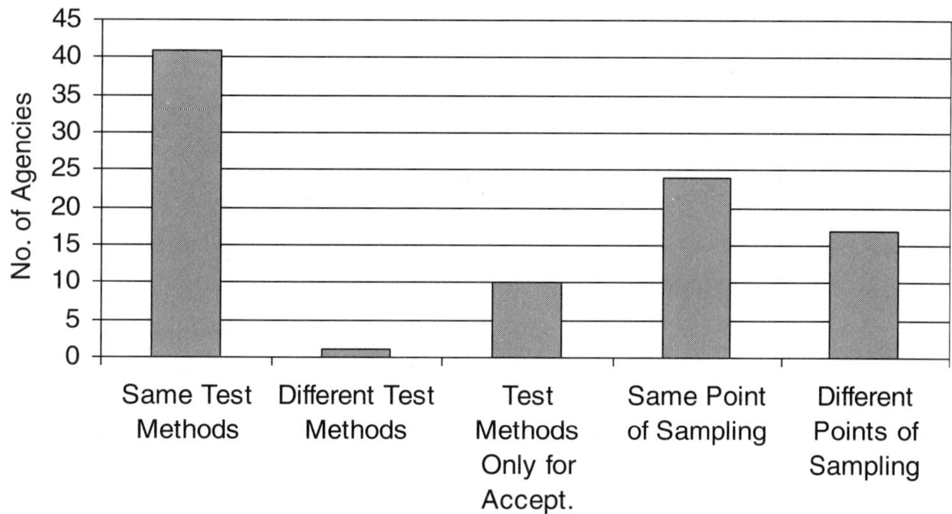

FIGURE 12 Test methods and point of sampling used for HMA (45 responses).

Thirty-three agencies accept HMA based on gradation, 26 use one or more volumetric properties for acceptance, 23 use fractured faces, and 22 use thickness as an acceptance attribute. Some of the lesser-used acceptance attributes are segregation, bulk specific gravity, and temperature.

Six agencies use an accept/reject acceptance plan, with 39 using a pay adjustment system. Of the 39 that use pay adjustment, 23 use a stepped pay schedule and 19 use equations (the numbers indicate that some agencies use more than one system depending on the attribute). Of those agencies that use equations, most were not identified. Of those that were identified, three use the equation Pay Factor = 55 + 0.5(PWL), one uses Pay Factor = 53 + 0.5(PWL), one uses Pay Factor = 83 + 0.2(PWL), and one uses a PWL value of 93 to determine whether an incentive is justified. No one single agency uses only an incentive, whereas 9 use only a disincentive, and 32 use both. Twenty-nine agencies use contractor test results as part of the acceptance decision. For procedures where the agency uses contractor test results in the acceptance decision, 20 agencies use all attributes, 3 use only attributes based on accept/reject, 2 use only attributes that do not involve pay, and 5 use attributes based on one or more of the following:

- Small qualities (i.e., less than 100 tons per day);
- Contractor tests for mix, agency tests for road; or
- A combination of volumetric properties and asphalt content.

The verification system based on one contractor test to one verification test is used by 11 agencies, with 2 using AASHTO D2S tolerances, 9 using agency-established tolerances, and 1 uses other verification procedures. Ten agencies use a comparison of accumulated tests; seven of these use the *F*- and *t*-tests, two use only the *t*-test, and one uses the AAD. Fourteen agencies use one agency test compared

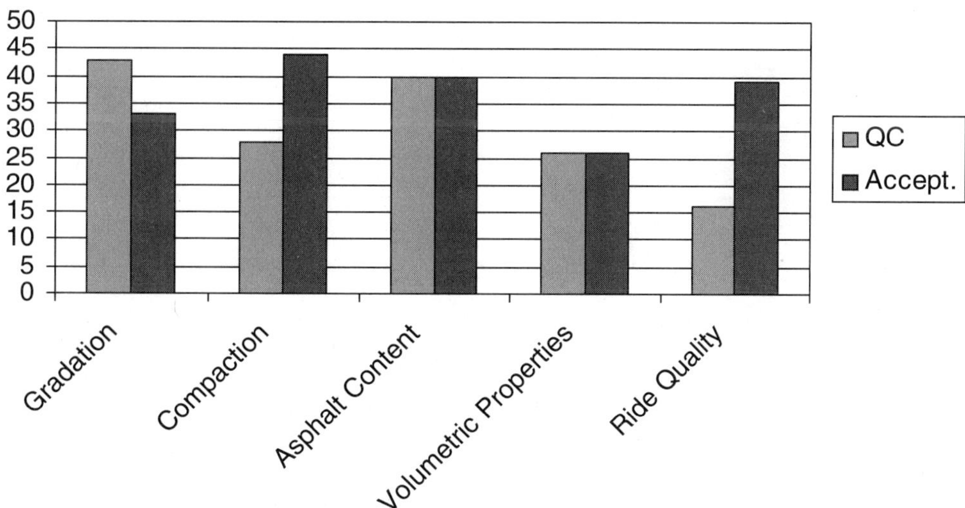

FIGURE 13 Attributes most often used for QC and acceptance of HMA (45 responses).

TABLE 5
ATTRIBUTES USED FOR QC AND ACCEPTANCE OF HMA

Attribute	QC	Acceptance
Asphalt content	40	40
Gradation	43	33
Compaction	28	44
Ride quality	16	39
Voids total mix	20	26
Voids in mineral aggregate	26	23
Aggregate fractured faces	25	23
Thickness	13	22
Voids filled with asphalt	19	13

Note: 44 responses.

TABLE 6
QUALITY MEASURES USED FOR HMA

Quality Measure	No. of Agencies
Percent within limits	26
Range	15
Average	13
Individual values	4
Average absolute deviation	4
Standard deviation	3
Percent defective	1
Moving average	1

Note: 45 responses.

with several contractor tests. Fourteen agencies use the comparison based on a lot and three use a complete project. Nine use independent samples, nine use split samples, and seven use both.

QUALITY MEASURES USED FOR ACCEPTANCE

Quality measures used for acceptance of HMA are shown in Table 6 and Figure 14. The most common quality measure for HMA, used by 27 agencies, is PWL or the complement PD. Fifteen agencies use the range, 13 agencies use the average, 4 use individual values, 4 use AAD, and 3 use the standard deviation.

TRAINING AND CERTIFICATION

Training and certification for agency personnel involved in inspection, sampling, and testing of HMA is mostly either in-house or in regional programs (see Figure 15). Thirty agencies have in-house training and 26 require certification. Ten agencies use regional programs for agency personnel training and 15 use regional certification. In addition, four agencies use joint agency–industry certification, and one each use National Asphalt Pavement Association (NAPA) training, university training, college training, and state board certification. For contractor personnel, 20 agencies use in-house training and 26 require in-house certification. Additionally, 9 agencies use regional training for contractor personnel, 15 require regional certification, 3 use joint agency–industry certification, and one each use university training, state board of registration, NAPA, and in-house contractor training.

CANADIAN QUALITY ASSURANCE PRACTICES

The Canadian QA practices for HMA differ from those used in the United States in a few instances. For types of QA programs, one province uses materials and methods provisions, another requires contractors to be ISO 9001-2000 certified, two use QA programs with the contractor controlling quality and agency performing acceptance, and one uses contractor test results in the acceptance decision. The QC and acceptance attributes used are generally based on gradation, asphalt con-

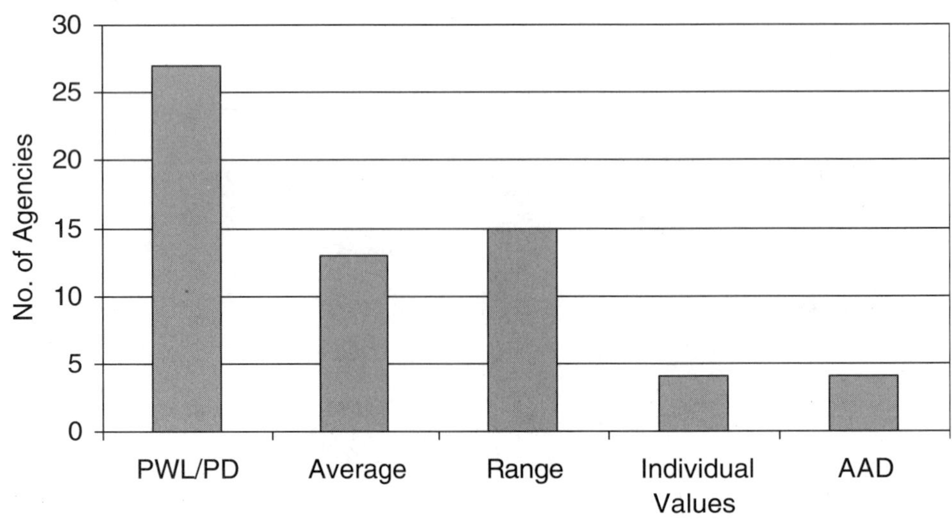

FIGURE 14 Quality measures most often used for acceptance of HMA (45 responses). PWL = percent within limits; PD = percent defective; AAD = average absolute deviation.

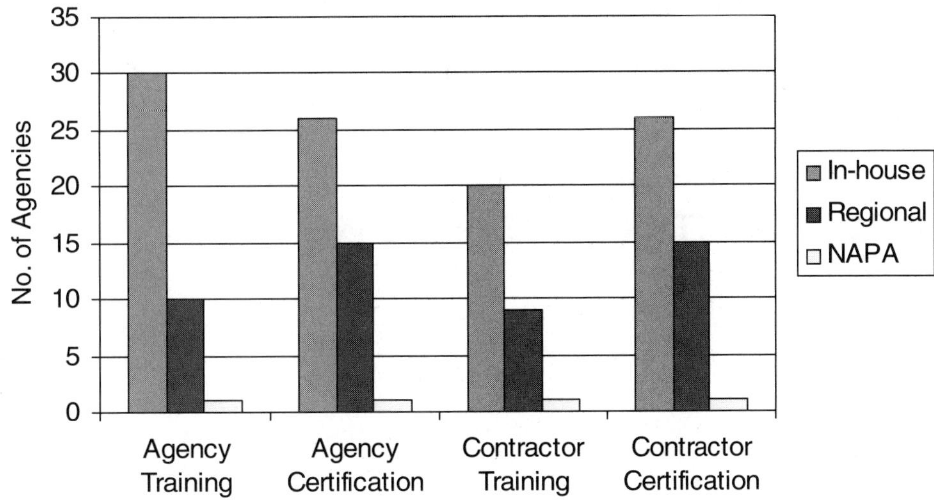

FIGURE 15 Training and certification requirements for HMA (45 responses). NAPA = National Asphalt Pavement Association.

tent, ride quality, and compaction. To a lesser extent, they use volumetric properties and thickness. Similar to the U.S. practices, they tend to use the same test methods and point of sampling for both QC and acceptance. For QC, four provinces use an agency-established frequency for QC tests and three require control charts. Two provinces use PWL, two use averages, two use range, and one uses individual values for acceptance. (The numbers indicate that some provinces use more than one quality measure depending on the attribute.) Of the five provinces responding, four use a pay adjustment system for acceptance; three of these use a stepped-pay schedule and one uses an equation. Two use only a disincentive and two use both an incentive and disincentive. Two provinces use contractor test results in the acceptance decision for all attributes; and they use either the *F*- and *t*-tests or a comparison of pay factors based on PWL for verification. One province compares tests on a lot basis and the other uses production over an extended period of time. One takes independent samples and the other takes split samples. Three provinces require in-house training of province personnel, one requires in-house training of contractor personnel, and one requires regional certification for contractor personnel.

CHAPTER SIX

QUALITY ASSURANCE PROGRAMS FOR PORTLAND CEMENT CONCRETE PAVING

Statistically based QA programs for PCC paving (PCCP) are not used by as many agencies as for HMA; however, such use has increased in the last decade. As an example, the Washington State DOT has developed QA specifications to resist the stresses at urban intersections (*53*). The use of performance-related specifications for PCCP is on the increase and appears to be ahead of the use of this type specification for HMA (*10,54*).

TYPE OF QUALITY ASSURANCE PROGRAM

Three agencies do not use PCCP and two agencies did not respond to this part of the questionnaire. Figure 16 shows that of the 40 responses to this material/construction area, 15 indicated that they use materials and methods provisions for PCCP, whereas 11 control the quality and perform acceptance. Sixteen agencies use QA programs with the contractor controlling quality and the agency performing acceptance, and 13 agencies use QA programs with the contractor controlling the quality and the agency using contractor test results in the acceptance decision.

Figure 17 shows that 32 agencies require the same test methods for QC and acceptance and that 24 use the same point of sampling. Two agencies require a different test method for QC and acceptance and eight agencies specify test methods only for acceptance. Eight use different points for sampling for QC and acceptance; of these, four use separate independent random locations for each, two allow the contractor to choose the point for QC sampling, and two have QC samples taken at the plant and sample for acceptance from the road.

QUALITY CONTROL

The attributes used for both QC and acceptance of PCCP are shown in Table 7 and Figure 18. Those attributes used most often for QC are air content, gradation, and slump. Twenty-five agencies use air content and gradation and 24 use slump. Additional often-used QC attributes are cylinder strength, thickness, beam strength, water–cement ratio, and aggregate fractured faces.

For the frequency of QC tests, 26 agencies use an agency-established frequency and 4 allow the contractor to choose the frequency. Seven agencies require the use of control charts.

ACCEPTANCE

The attributes used most often for acceptance of PCCP are air content, which is used by 38 agencies and thickness used by 36 (see Table 7 and Figure 18). Thirty-three agencies accept PCCP based on slump, 31 agencies accept PCCP based on cylinder strength, 26 use gradation, 18 use beam strength, 16 use water–cement ratio, and 15 use ride quality. Some of

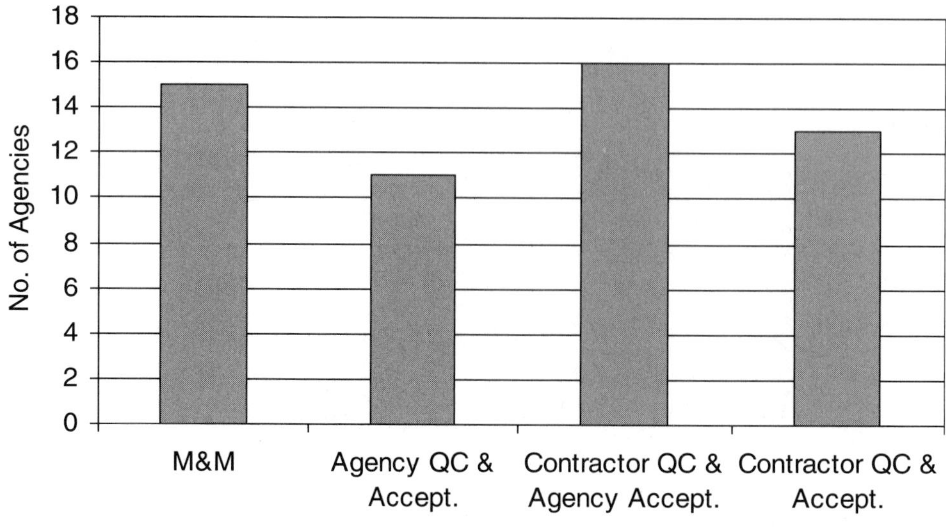

FIGURE 16 QA programs for PCC paving (40 responses). M&M = materials and methods.

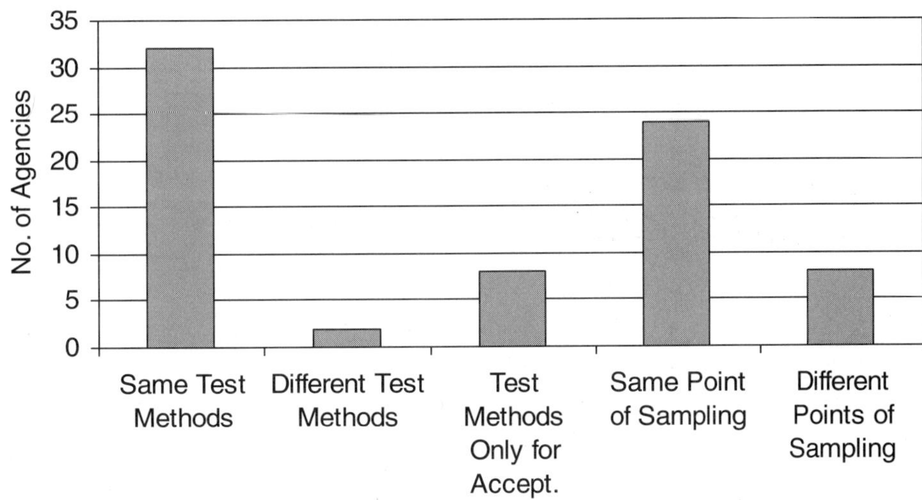

FIGURE 17 Test methods and point of sampling for PCCP (40 responses).

the lesser-used acceptance attributes are aggregate fractured faces, sand equivalence, core strength, permeability, and surface tolerance and texture.

Seventeen agencies use accept/reject acceptance plans and 28 use a pay adjustment system. Of the 28 that use pay adjustment, 21 use stepped pay schedules, 7 use equations, and 4 use a combination of methods. One uses only an incentive, 12 use only a disincentive, and 16 use both. Fourteen agencies use contractor test results as part of the acceptance decision. For procedures where the agency uses contractor test results in the acceptance decision, eight agencies use all attributes, two use only attributes that do not involve pay, and three use other attributes.

The verification system based on one contractor test to one verification test is used by six agencies; of those, one uses AASHTO D2S, two use ASTM D2S tolerances, two use agency-established tolerances, and two use other verification procedures. Three agencies use *F*- and *t*-tests for a comparison of accumulated tests, and one uses only the *t*-test. Five agencies use one agency test compared with several contractor tests. Four agencies use a comparison of accumulated test results based on a lot, four use results based on a completed project, and three use a production over an extended period of time. Five use independent samples, four use split samples, and four use both.

QUALITY MEASURES USED FOR ACCEPTANCE

Table 8 and Figure 19 show the quality measures used for acceptance of PCCP. The most commonly used quality measure, PWL/PD, is used by 16 agencies, with the range next being used by 15 agencies. Twelve agencies use the average,

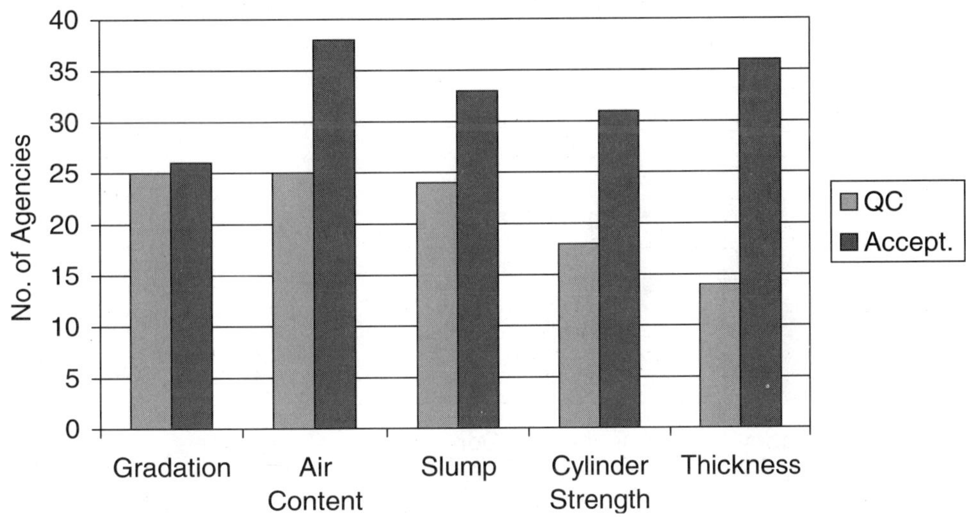

FIGURE 18 Attributes used most often for QC and acceptance of PCCP (40 responses).

TABLE 7
ATTRIBUTES USED FOR QC AND ACCEPTANCE OF PCCP

Attribute	QC	Acceptance
Air content	25	38
Thickness	14	36
Slump	24	33
Cylinder strength	18	31
Gradation	25	26
Beam strength	14	18
Water–cement ratio	12	16
Ride quality	1	15
Aggregate fractured faces	7	6
Sand equivalence	0	3
Permeability	0	3
Core strength	0	2

Note: 40 responses.

TABLE 8
QUALITY MEASURES USED FOR PCCP

Quality Measure	No. of Agencies
Percent within limits	13
Range	15
Average	12
Individual values	10
Standard deviation	3
Percent defective	3

Note: 40 responses.

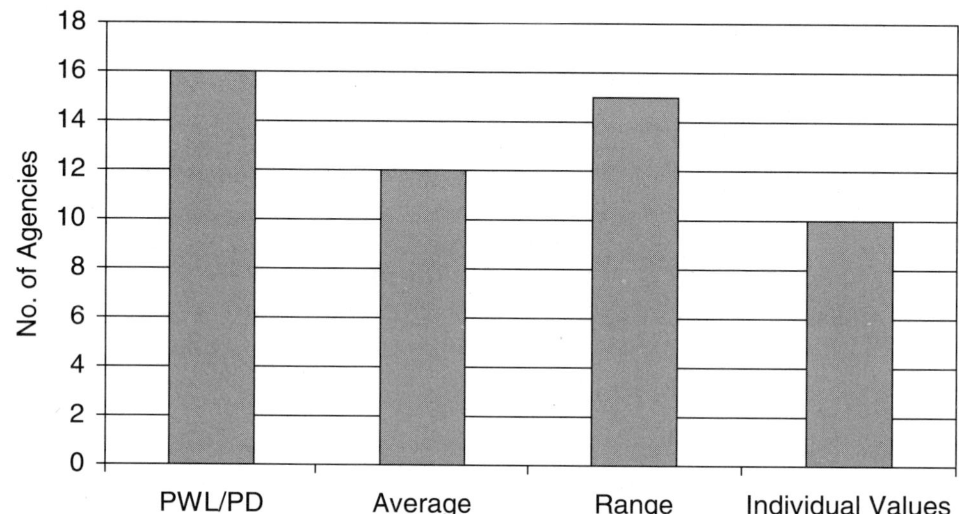

FIGURE 19 Quality measures most often used for PCCP (40 responses). PWL = percent within limits; PD = percent defective.

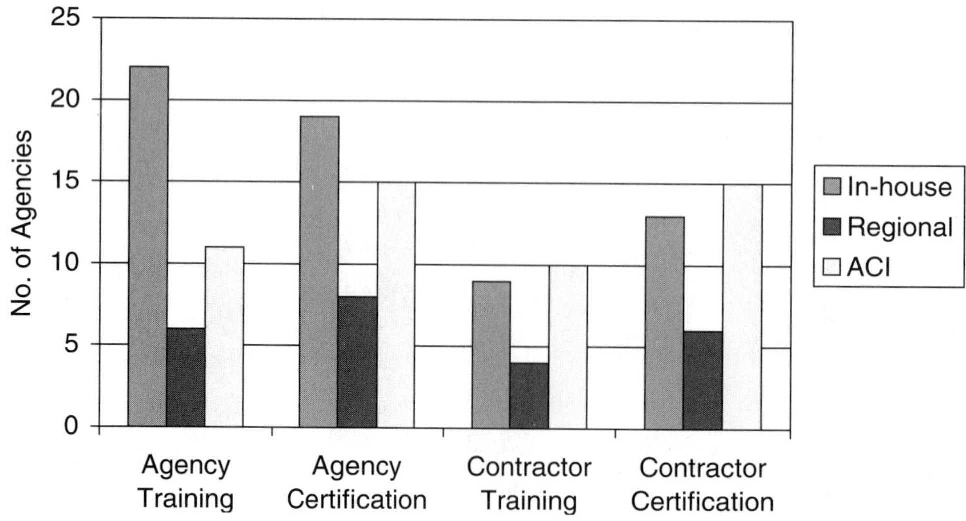

FIGURE 20 Training and certification requirements most often used for PCCP (40 responses). ACI = American Concrete Institute.

10 use individual values, and 3 use standard deviation. There are multiple combinations of the use of the range, the average, and the standard deviation. Also, depending on the attribute, sometimes these quality measures are used by themselves.

TRAINING AND CERTIFICATION

Training and certification for agency personnel involved in inspection, sampling, and testing of PCCP uses mostly in-house, American Concrete Institute (ACI), or regional programs. Figure 20 shows that 22 agencies have in-house training and 19 require in-house certification for agency personnel. Eleven agencies use ACI for training, and 15 use certification from this source. Six agencies use regional programs for agency personnel training and eight use regional certification. Other training and certification programs used by one responding agency include university training, college training, the American Concrete Paving Association (ACPA), Precast/Prestressed Concrete Institute (PCI), and state board certification. For contractor personnel, 9 agencies require in-house training and 13 require in-house certification. Additionally, 10 agencies use ACI for training and 15 for certification, 4 allow contractor personnel to receive regional training and 6 use regional certification. Other training and certification programs used include university training, college training, ACPA, or state board of registration.

CANADIAN QUALITY ASSURANCE PRACTICES

The Canadian QA practices for PCCP primarily use QA programs with the contractor controlling quality and the agency performing acceptance. One province uses materials and methods provisions. The QC and acceptance attributes used are generally based on cylinder strength, slump, thickness, and ride quality. To a lesser extent, the responding provinces use beam strength and air content. Similar to the U.S practices, they tend to use the same test methods and point of sampling for both QC and acceptance. Two provinces use an agency-established frequency for QC tests. Two provinces use PWL, two use the range, and one uses averages for acceptance. One province uses accept/reject/rework provisions and three use pay adjustment systems for acceptance. One province uses contractor tests in the acceptance decision for all attributes. This province takes both independent and split samples. Agency personnel training is required by two provinces using in-house programs, whereas two use ACI for certification. One agency requires either an in-house program or ACI certification for contractor personnel. The Canadian Standards Association administers one of the training programs used by two provinces for agency and contractor personnel.

CHAPTER SEVEN

QUALITY ASSURANCE PROGRAMS FOR PORTLAND CEMENT CONCRETE STRUCTURES

As expected, the QA programs for PCC structures are often similar to those of PCCP. Therefore, statistically based QA programs for PCC structures are not used by as many agencies as those for HMA; however, the use is increasing in this area (*55*).

TYPE OF QUALITY ASSURANCE PROGRAM

Figure 21 shows that of the 43 respondents, 25 use materials and methods provisions for PCC structures, with 14 agencies controlling the quality and performing acceptance. Seventeen agencies use QA programs with the contractor controlling quality and the agency performing acceptance and 13 agencies use QA programs with the contractor controlling the quality and the agency using the contractor test results in the acceptance decision.

Figure 22 shows that 36 agencies require the same test methods for QC and acceptance and 26 use the same point of sampling. One agency requires a different test method for QC and acceptance and 10 agencies specify test methods only for acceptance. Six use different points for sampling for QC and acceptance; five using different independent random samples for QC and acceptance and one samples for QC from the beginning of truck discharge and samples for acceptance from the middle of discharge.

QUALITY CONTROL

The attributes used for both QC and acceptance of PCC structures are shown in Table 9 and Figure 23. The attributes used most often for QC are gradation, slump, air content, and cylinder strength. Thirty agencies use gradation, 29 use slump, 28 use air content, and 21 use cylinder strength. Another often-used QC attribute is water–cement ratio, which is used by 15 agencies. Lesser-used attributes for QC are aggregate fractured faces and permeability.

For the frequency of QC tests, 30 agencies use an agency-established frequency and 7 let the contractor choose the frequency. Five agencies require the use of control charts.

ACCEPTANCE

Table 9 and Figure 23 show that the attributes used most often for acceptance of PCC structures are air content, used by 42 agencies; cylinder strength and slump, each used by 40; and gradation used by 30. Seventeen agencies accept PCC structures based on water–cement ratio, and 11 agencies accept PCC structures based on aggregate fractured faces. Some of the lesser-used acceptance attributes are permeability, temperature, sand equivalence, and beam strength.

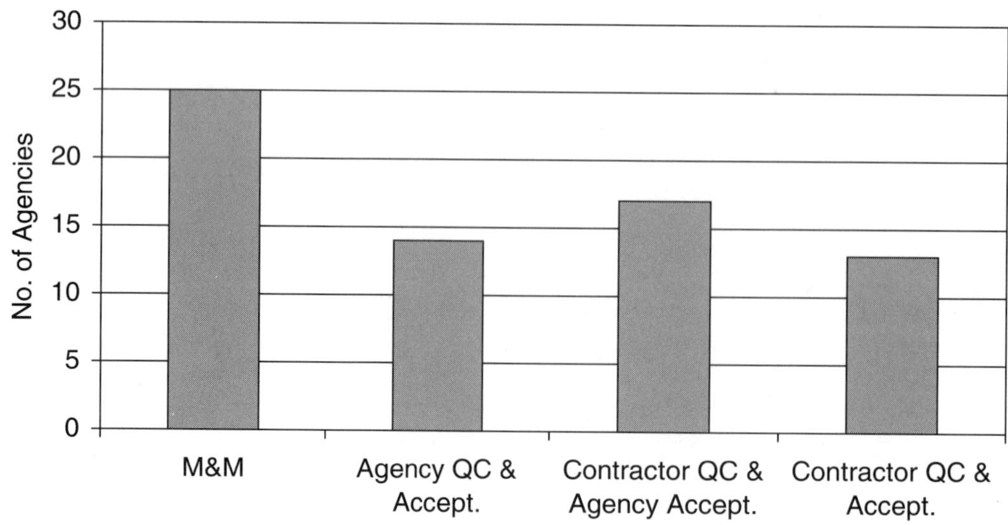

FIGURE 21 QA programs for PCC structures (43 responses). M&M = materials and methods.

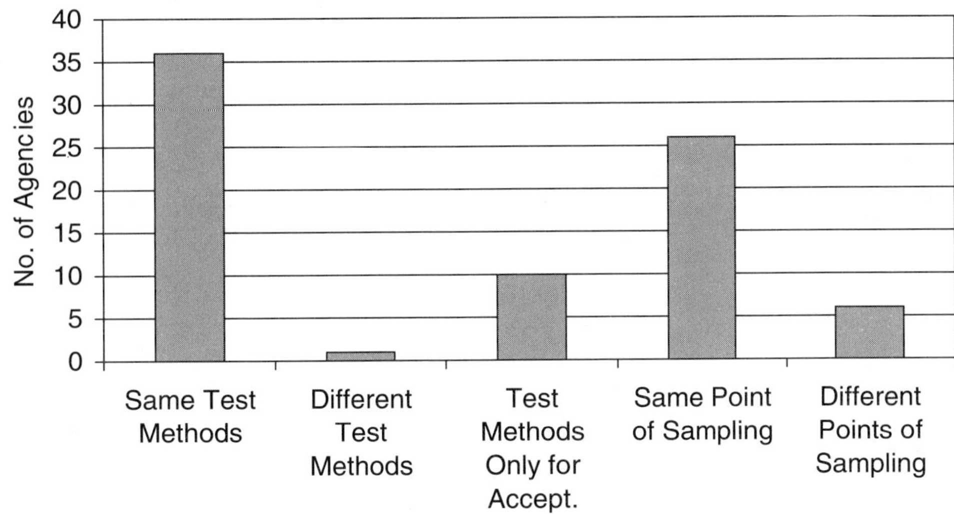

FIGURE 22 Test methods and point of sampling for PCC structures (43 responses).

Nineteen agencies use accept/reject acceptance plans and 28 use pay adjustment systems. Of the 28 that use pay adjustment, 14 use a stepped pay schedule and 10 use equations. No one single agency uses only an incentive, whereas 19 use only a disincentive, and 9 use both. Fourteen agencies use contractor test results as part of the acceptance decision. For procedures where the contractor test results are used in the acceptance decision, 10 agencies use all attributes, 1 agency uses only attributes based on accept/reject, and 2 use only attributes that do not involve pay.

Seven agencies use a verification system based on one contractor test to one verification test. Of those, one uses AASHTO D2S, two use ASTM D2S tolerances, two use agency-established tolerances, and two use other verification procedures. This number indicates that some agencies use more than one source. One agency uses F- and t-tests for a comparison of accumulated tests and one uses only the t-test. Ten agencies use one agency test compared with several contractor tests. Three agencies use a comparison of accumulated test results based on a lot, three use a completed project, and two use production over an extended period of time. Two use independent samples, four use split samples, and seven use both.

QUALITY MEASURES USED FOR ACCEPTANCE

Table 10 and Figure 24 show the quality measures used for acceptance of PCC structures. The most common quality mea-

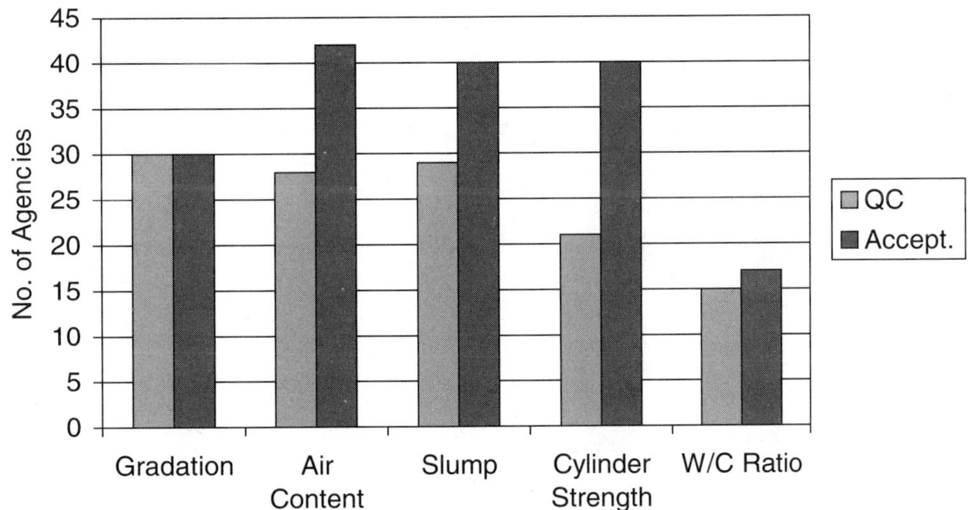

FIGURE 23 Attributes used most often for QC and acceptance of PCC structures (43 responses). W/C Ratio = water–cement ratio.

TABLE 9
ATTRIBUTES USED FOR QC AND ACCEPTANCE OF PCC STRUCTURES

Attribute	QC	Acceptance
Air content	28	42
Cylinder strength	21	40
Slump	29	40
Gradation	30	30
Water–cement ratio	15	17
Aggregate fractured faces	7	11
Permeability	5	8
Temperature	1	2
Sand equivalence	0	2
Beam strength	0	2

Note: 43 responses.

TABLE 10
QUALITY MEASURES USED FOR PCC STRUCTURES

Quality Measure	No. of Agencies
Average	16
Range	16
Percent within limits	8
Individual values	8
Standard deviation	6
Percent defective	2

Note: 43 responses.

sures, used by 16 agencies are the average and range, 10 use the PWL or PD, 8 use individual values, and 6 use the standard deviation.

TRAINING AND CERTIFICATION

Figure 25 shows that training and certification for agency personnel involved in inspection, sampling, and testing of PCC structures uses mostly in-house, ACI, or regional programs. Twenty-five agencies require in-house training and 24 require in-house certification. Eleven agencies use ACI for training and 16 use certification from this source. Six agencies use regional programs for agency personnel training and 10 use regional certification. University, college, ACPA, and PCI training and certification, and state board certification are also used. For contractor personnel, 11 agencies require in-house training and 14 require in-house certification. Also, 12 agencies use ACI for training and 15 for certification, 4 allow contractor personnel to receive regional training, and 7 use regional certification. University, college, ACPA, and PCI training and certification, and state board certification are also used for contractor personnel.

CANADIAN QUALITY ASSURANCE PRACTICES

The Canadian QA practices for PCC structures primarily use QA programs, with the contractor controlling quality and the agency performing acceptance. Two provinces use materials and methods provisions. One uses the contractor controlling the quality and contractor tests used in the acceptance decision. The QC and acceptance attributes used are generally based on cylinder strength and slump. Used to a lesser extent are permeability, aggregate fractured faces, and air content. Similar to U.S. practices, the responding provinces tend to use the same test methods and point of sampling for both QC and acceptance. Three provinces use an agency-established frequency for QC tests, with three provinces

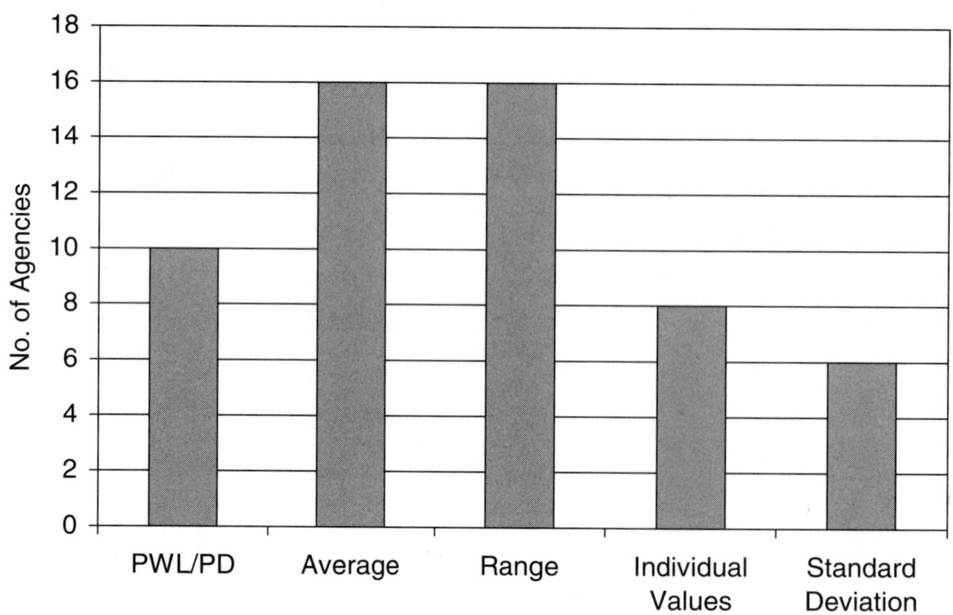

FIGURE 24 Quality measures most often used for PCC structures (43 responses). PWL/PD = percent within limits/percent defective.

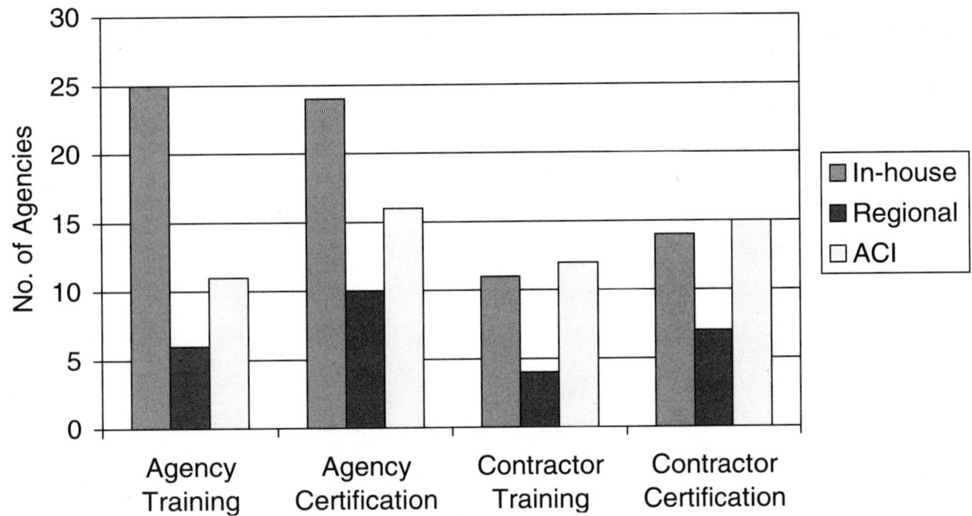

FIGURE 25 Training and certification requirements most often used for PCC structures (43 responses). ACI = American Concrete Institute.

using PWL, one using the range, and one using averages for acceptance. Two provinces use accept/reject/rework provisions and three use pay adjustment systems for acceptance. One province uses contractor test results in the acceptance decision based on air voids in the hardened concrete. For verification, this province compares one agency test to several contractor tests. This province takes split samples. Agency personnel training is required by two provinces using in-house programs and two use ACI for certification. One agency requires either in-house program or ACI certification for contractor personnel. The Canadian Standards Association administers a training program used by three provinces for agency personnel and two use it for contractor personnel.

CHAPTER EIGHT

INDEPENDENT ASSURANCE

As reported in chapter two, IA is used by agencies in two different contexts. Some use IA in the narrower context, to provide an independent assessment of the test results obtained from QC and acceptance. Others use the function in the broader context, to provide an independent assessment of the product and/or the reliability of test results obtained from the process control and acceptance testing. As an independent evaluation, it is designed to provide a complement to the QC and acceptance functions and, in doing so, it involves a separate and distinct schedule of sampling, testing, and observation.

The survey questionnaire attempted to determine how the IA unit in each agency is organized, how many FTE personnel are used in the IA function performing different tasks, and to what testing the agency applies the IA function. The responses indicated that the administration of the IA system varies appreciably from agency to agency. In some agencies, the IA personnel have only the IA function to conduct. In others, the IA function is only one of several that they perform.

INDEPENDENT ASSURANCE STAFFING ORGANIZATION AND APPLICATION

Figure 26 shows how agencies responded to the question "How is IA organized in your agency?" Twenty-eight agencies are organized statewide, 15 by district or region, 10 by project, and 10 by system. Response comments clarified that some agencies use a nested approach that can be included under more than one category.

Figure 27 shows the source to which IA testing is applied. Forty-two agencies apply the IA function to agency testing, 26 to contractor testing, 15 to producer testing, 10 to supplier testing, and 3 to consultant testing.

INDEPENDENT ASSURANCE STAFFING BY FULL-TIME EQUIVALENTS

As mentioned previously, the IA staffing varies greatly from agency to agency; for example, one agency has approximately 3,000 technicians that perform IA sampling and/or testing as part of their work function, whereas another agency has just four technicians that perform IA sampling and testing as their sole function. Of the 45 respondents, 31 responded to the IA questions on the FTEs and budget, and 29 of these produced the data needed to analyze this function appropriately. In addition to examining the IA FTEs by agency, an analysis was made by using the agency's construction and maintenance budget as a normalizing factor to the FTE.

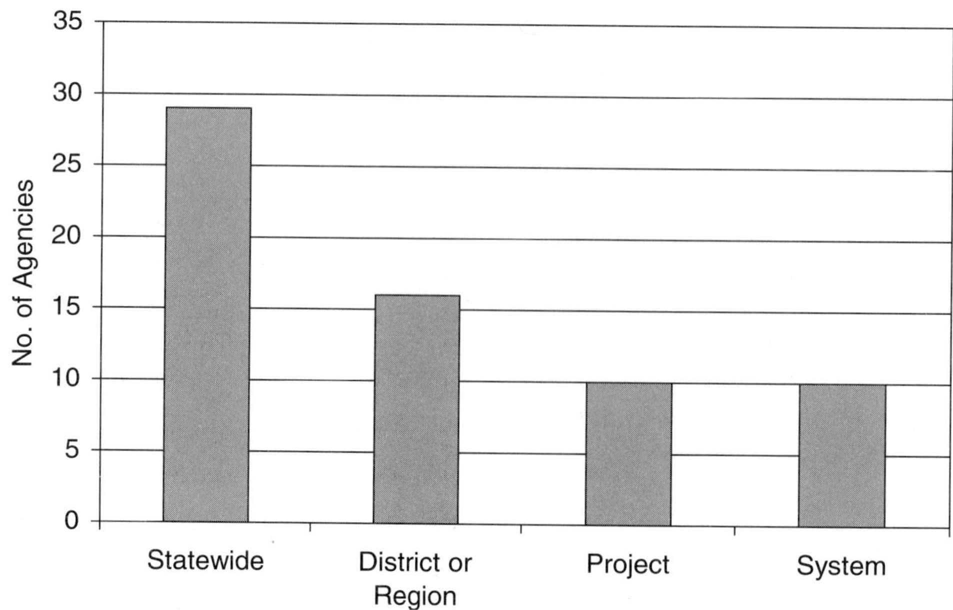

FIGURE 26 Organization of agency IA testing (45 responses).

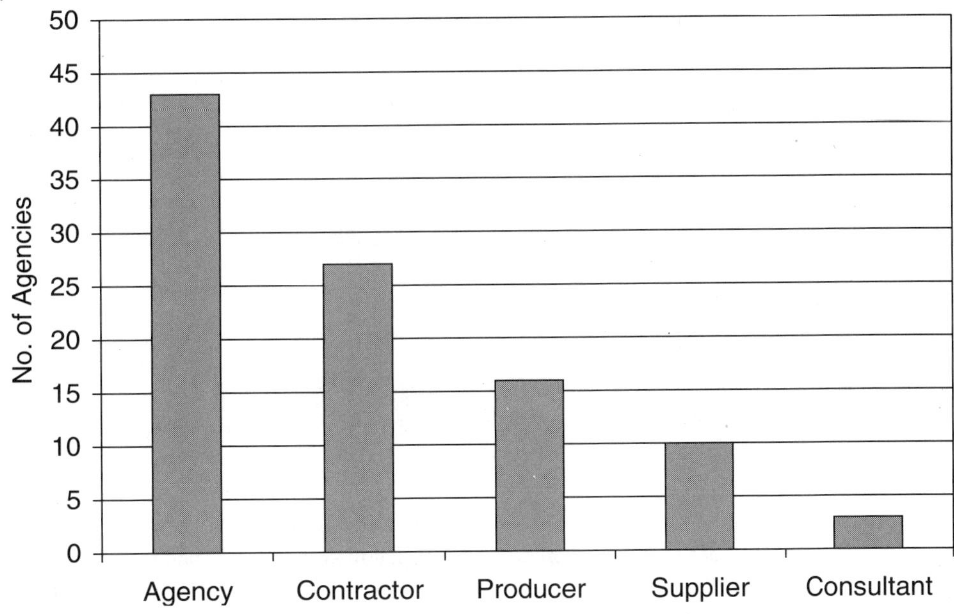

FIGURE 27 Application of IA testing (45 responses).

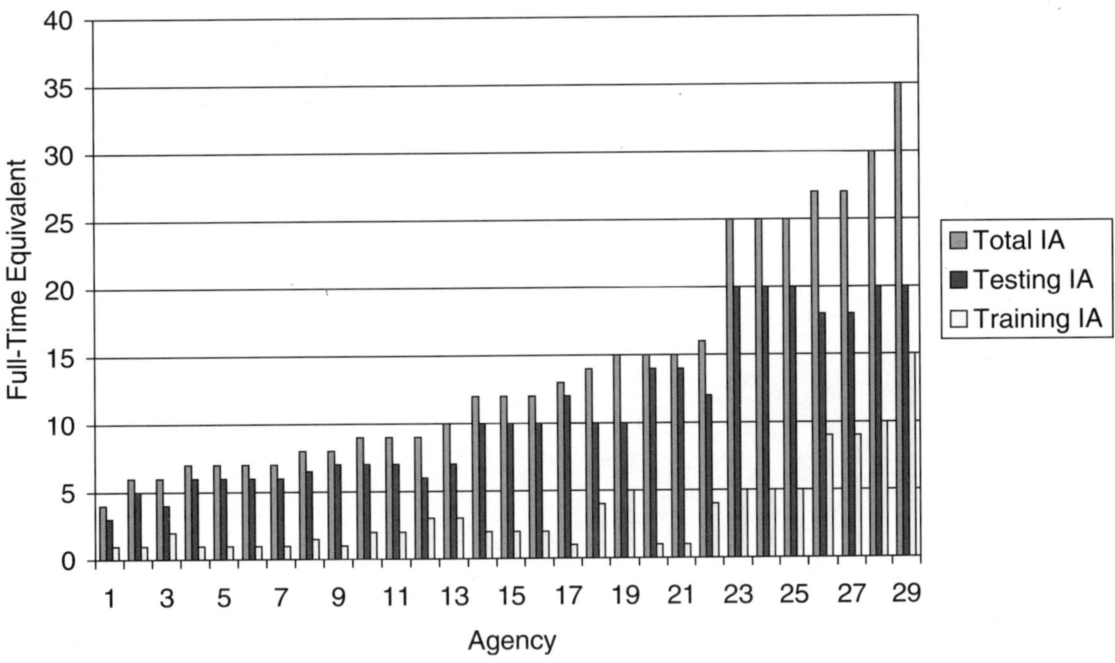

FIGURE 28 Total, sampling and testing, and training IA FTE staffing by agency.

Figure 28 shows the IA FTEs by agency. As shown, the number of total FTEs in an agency varies from 4 to 35. The number that perform sampling and testing varies from 3 to 20, and the number that perform training and recordkeeping varies from 1 to 15.

Using the agency's construction and maintenance budget as a normalizing factor produced Figure 29. From this figure it can be seen that total IA staffing varies from 0.5 FTE per hundred million dollars to 16 FTEs per hundred million dollars. The FTEs that perform sampling and testing vary from less than 0.5 to 12 per hundred million dollars, and the number that perform training and recordkeeping varies from less than 0.1 to 4 FTEs per hundred million dollars.

CANADIAN QUALITY ASSURANCE PROGRAMS

The Canadian provinces are not required to conduct IA testing under the same requirements as those used in the United States. However, all five provinces responding to the ques-

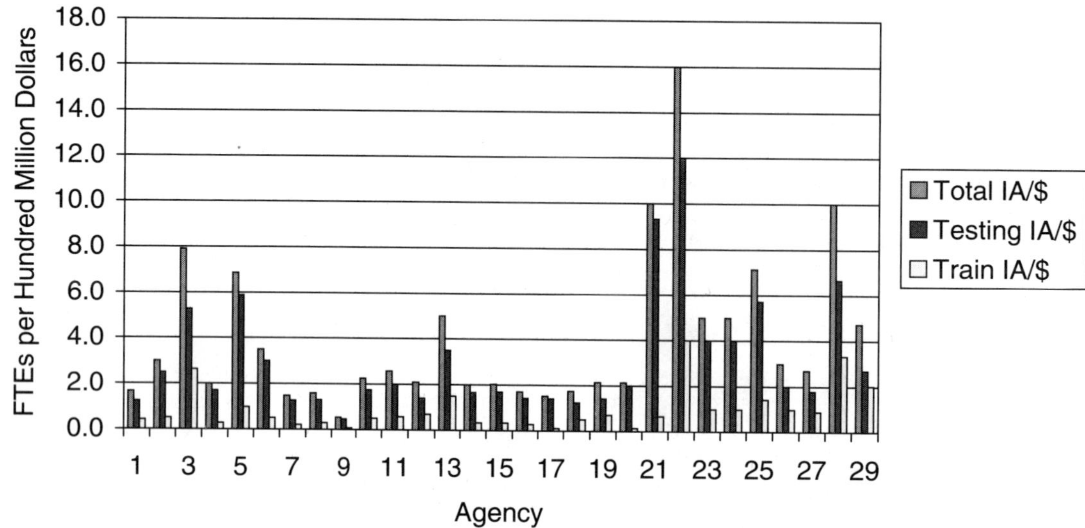

FIGURE 29 IA FTE staffing per hundred million dollar construction and maintenance budget by agency.

tionnaire indicated that they do use a form of IA testing using one of two approaches; three use a province-wide approach and the other two use a system-based approach. Three provinces apply IA testing to agency tests, one uses IA for producer testing, and one uses IA for supplier testing. Two provinces reported on their FTEs and construction and maintenance budgets. The IA staffing is relatively high compared with the United States. One province has a total of 12 IA FTEs and the other 15. Normalized by budgets, this comes out to be 20 and 15 FTEs per hundred million (Canadian) dollars.

CHAPTER NINE

CONSULTANTS, INNOVATIVE PRACTICES, AND FUTURE OF QUALITY ASSURANCE PROGRAMS

USE OF CONSULTANTS

Hiring outside consultants to perform QA functions is a common practice. *NCHRP Synthesis of Highway Practice 263*, published in 1998, reported that 17 agencies of 39 responding (44%) indicated that they contracted some QA testing outside of their workforce (*9*). These 17 agencies used consultants because of staff reductions and lack of personnel (10 responses), lack of qualified personnel (3 responses), and balancing the workload (4 responses). The survey questionnaire for this synthesis indicated that the number of agencies using consultants has increased to 35 out of the 45 U.S. agencies responding (78%). These 35 agencies use consultants for the following activities:

- In place of agency acceptance testing (23),
- As a supplement to agency acceptance tests (20),
- In place of contractor QC tests (12),
- In place of contractor tests used in the acceptance decision (8),
- As a supplement to contractor QC tests (6), and
- As a supplement to contractor tests used in the acceptance decision (2).

Of the 35 agencies that reported that they use consultants, 27 replied that the consultants were used to test the products shown in Table 11 (8 did not list the products.) Thirteen agencies employ consultants to test all products. The remaining 14 use consultants to test a combination of products, as shown in Table 11.

WARRANTIES

Warranties have become more commonplace for some agencies (*35,36*). However, the questionnaire responses indicated that only eight agencies routinely use them. This is the same number of agencies that responded that they were routinely using warranties in the 1998 synthesis survey (*9*). Four of the eight noted that they use warranties in the same manner as materials and construction used in their normal QA programs. This is interpreted as being a generalization, because they likely do not use the same level of QC, acceptance, and IA testing as in their normal QA programs. Of the four that treat them differently, two reported that they use no agency testing with warranty specifications, one uses them with materials and methods specifications on capital preventive maintenance contracts, and one uses them for seal coat maintenance.

OTHER INNOVATIVE PRACTICES

The use of other innovative practices has also generated a considerable amount of attention (*35,36*). Once again, however, the questionnaire responses indicated that only a few agencies routinely use them. Although these are different forms of contracting than the typical materials and methods or QA types, they still involve the control and acceptance of the materials and construction. This synthesis was interested in determining how agencies that use these newer procedures conduct QC and acceptance functions. The responses often did not answer this question. However, the responses did reveal that nine agencies routinely use some form of innovative practice. Of these nine, seven use design–build. One agency has used design–build on a major bridge and on a major urban road connector. Although not specific to QA programs, other innovative practices that were listed by the agencies as having been used routinely are:

- Lane rental,
- Community relations,
- Toll roads, and
- Private/public partnerships.

ACCEPTANCE BY CERTIFICATION

Different interpretations were applied to the question, "Do you accept any pavement materials solely by certification?" Some agencies apparently interpreted the question to be the finished product; that is, HMA, PCCP, etc. Others interpreted the question to apply to ingredient materials. Thus, the responses were somewhat mixed. However, 11 agencies responded that they accept the following products solely by certification:

- Five agencies accept admixtures,
- Three agencies accept reinforcing steel,
- Two agencies accept cement and fly ash,
- Two agencies accept small quantities (the quantities were not identified), and
- Two agencies accept noncritical items (not identified).

As the total of 14 indicates, some agencies accept more than one of the materials in the categories listed.

TABLE 11
PRODUCTS TESTED BY CONSULTANTS

Products Tested	No. of Agencies
All	13
PCC	8
HMA	5
Prestressed and precast PCC	4
Structural steel	3
Soils	3
Aggregates	2

Notes: 27 responses. PCC = portland cement concrete; HMA = hot-mix asphalt.

TABLE 12
PRODUCTS IN QA PROGRAMS WHERE CHANGES ARE EXPECTED

Product	No. Agencies
Paving and or/PCC structures	11
HMA mix and/or binder	10
Soils, embankments, and/or base courses	9
System	3
All	5
Manufactured products	2
Precast and prestressed concrete	1
Pipe	1
Reinforcing steel	1

Notes: 23 responses. PCC = portland cement concrete; HMA = hot-mix asphalt.

FUTURE OF QUALITY ASSURANCE PROGRAMS

Because QA programs are evolutionary, it was of interest to find out what changes were anticipated by the agencies. The first part of the question asked was "Do you anticipate significant changes in your QA program for any products in the near future?" and the second part asked for information on future directions. Twenty-three agencies reported that they anticipate significant changes in their QA programs in the near future and 22 indicated they did not anticipate any changes. Table 12 shows the products where changes are expected. The types of changes are discussed in Table 13. For completeness, all comments to the second part of this question are listed in Table 13. An important addition to changes in QA programs is FHWA's recently updated QA technical advisory directive (*56*).

CANADIAN QUALITY ASSURANCE PROGRAMS

Four of the Canadian provinces use consultants in place of or as a supplement to agency acceptance tests. None use warranties routinely. One province uses design–build and none accept materials by certification. No province anticipated significant changes in their QA programs in the near future.

TABLE 13
EXPECTED CHANGES IN QA PROGRAMS BY AGENCY

Agency	Comment
Arizona	Implementing third generation computerized workmanship program to be used by IA, agency, and contractor personnel.
California	(1) For manufactured materials; when the department implements a materials management system, the department will no longer perform QA on a project-by-project basis, but will release material on a manufacturer's track record (*57*). (2) Implement requirement for contractors to develop a QC plan with minimum acceptable frequency and observations including identification of a quality manager. Department QA will be "Did they follow the plan?" and perform statistically valid random sampling and separate tests. (3) Implement one-year workmanship and warrantee program. (4) Plan to aggressively move to performance/end-result specifications.
Colorado	Will implement move to using contractor's test results as part of the acceptance decision for HMA when Colorado DOT acceptance is based on voids.
Delaware	Based on success of recently implemented HMA QA program; will develop acceptance based on prorated payment for PCC, soils, and aggregates.
Georgia	Planning to develop a system-based IA program.
Idaho	QA specifications currently under development for PCCP and PCC structures.
Kansas	Develop QA program for soils similar to ones for PCCP and HMA.
Kentucky	Moving toward "Total Project QC" by 2005. Contractor will be required to have qualified individuals to cover all disciplines. Agency will perform verification and use contractor test results as part of the acceptance decision.
Louisiana	(1) Use "in-place" penetration tests for acceptance of base course. (2) Use concrete beams for flexural strength instead of compressive strength tests on cores for PCCP. (3) Use contractor surface tolerance test results as part of the acceptance decision for ride quality.
Maine	(1) Develop statistical acceptance for PCC structures and recycled base. (2) Looking into performance-related specifications for HMA.
Massachusetts	Develop regional QA program for acceptance of manufactured products.

TABLE 13 (*Continued*)

Agency	Comment
Minnesota	Develop QA program for soils and base.
Montana	(1) Develop QA program using contractor test results in the acceptance decision for HMA, PCCP, and aggregate surfaces. (2) Develop a qualified products list.
Nevada	Develop a computerized test reporting system.
New Hampshire	Develop QA contract provisions for base course materials and construction.
New Mexico	Use Aggregate Index for acceptance of PCCP and volumetric properties for acceptance of HMA.
New York	Reduce number of materials requiring stock lot sampling and testing for acceptance. Developing a comprehensive QA procedure to reduce or eliminate the need for individual methods/procedures for each material.
North Carolina	Develop statistically based specifications and/or performance-related specifications for PCCP and HMA.
Ohio	Moving to American Concrete Institute design, contractor control, and agency verification for PCCP and PCC structures.
Pennsylvania	Changing the way HMA is accepted.
South Carolina	Develop QA program using contractor/producer test results in the acceptance decision for PCCP and structural steel.
Texas	(1) Implementing seismic methods and dielectric testing as part of the performance quality management of soils and embankments. (2) Contractors will be required to have QC personnel with ACI, Precast/Prestressed Concrete Institute (PCI), or other recognized certification for producing precast and prestressed concrete bridge beams. The contractor test results will be used in the acceptance decision with agency verification testing at a reduced rate.

Note: 23 responses.

CHAPTER TEN

CONCLUSIONS AND FUTURE RESEARCH NEEDS

This synthesis addresses the types of quality assurance (QA) programs used by state, federal, and Canadian transportation agencies for control and acceptance of typical highway pavement materials and construction. One of the first subjects discussed in the synthesis was establishing the commonly agreed on definitions used for QA programs. A problem with QA programs that has existed since their inception has been different interpretations of the terms. Fortunately, a valid glossary of highway QA terms was available from TRB and proved to be a great asset.

The following general conclusions are drawn from the literature reviewed and from responses to the survey questionnaire of the state of the practice of QA programs.

As expected, the state QA programs are quite varied. Some types of programs are used more frequently than others within a material type; for example, materials and methods for soils and embankments, and agency use of contractor test results in the acceptance decision for hot-mix asphalt (HMA). Therefore, although some definite trends were reported, each agency generally has its own ideas and reasons as to how and why its QA program operates as it does. An example of the diversity can be found in the use of pay adjustment schedules. Most of the agencies use stepped pay schedules that, naturally, vary appreciably from state to state. Some agencies use pay equations, and for those no single equation is used appreciably more than another. One area of relative agreement is in training and certification. Most agencies tend to rely on in-house training and certification when either is required for all material and construction areas. For example, more than 70% of those responding use in-house training for agency personnel testing for soils and embankments and for base and subbase and approximately 50% require in-house certification for their personnel for these materials.

Two practices were found that indicated that QA programs were not being used to their optimum capability. The first is that although the concept of QA calls for the separation of the functions of quality control (QC) and acceptance, it is not clear that the majority of agencies clearly separate them. For all material and construction areas surveyed, both QC and acceptance functions often overlap by the use of the same test methods and point of sampling, irrespective of whether the agency or the contractor performs the test. The second practice was also apparent from the questionnaire responses. An appreciable number of agencies use simpler but statistically weaker procedures for the type of verification system when the agency uses the contractor test results as part of the acceptance decision. This procedure is less sensitive in measuring differences between agency and contractor test results.

The general QA practices by material and construction area are recapped here.

- QA programs for soils and embankments use more materials and methods specifications than are used for other materials. Moisture content and compaction are the two most often used attributes for both QC and acceptance. All respondents use accept/reject provisions as opposed to pay adjustment. The quality measure used most often for acceptance is individual values; confirming earlier conclusions that the variability of soils and embankments makes the use of more powerful statistical analysis tools (in conjunction with more sampling and testing) less practical.
- QA programs for aggregate base and subbase are fairly evenly divided between the agency controlling the quality and performing acceptance and the contractor controlling the quality and the agency performing acceptance. More agencies use contractor tests in the acceptance decision for this material area than use this type of program for soils and aggregates. Gradation, moisture content, and compaction are the most often-used attributes for both QC and acceptance. Aggregate fractured faces percentage is also an often-used acceptance attribute. More than three-quarters of the respondents use accept/reject provisions and approximately one-quarter use pay adjustment procedures. Eight of those using pay adjustments tend to use stepped pay schedules; 11 use only a disincentive and 5 use both an incentive and disincentive. The quality measure used most often for acceptance is individual values, but a sizeable number use percent within limits (PWL) [or percent defective (PD)] (13) and range (13). Of the 13 agencies that use contractor test results in the acceptance decision, six compare one contractor test with one agency test, five use the *F*- and *t*-tests, and three use one agency test compared with several contractor tests. For verification, five use independent samples, two use split samples, and three use both.
- More than half of the agencies use QA programs for HMA, in which the contractor controls the quality and the agency uses contractor test results in the acceptance

decision. A substantial number use the practice of the contractor controlling the quality and the agency performing acceptance; few use materials and methods-type provisions. Gradation, asphalt content, volumetric properties, and compaction are the most frequently used QC attributes. Three-quarters of the agencies establish the frequency for contractors to conduct QC tests, and 27 agencies require control charts. Gradation, asphalt content, volumetric properties, compaction, and ride quality are the attributes most often used for acceptance. Aggregate fractured faces percentage and thickness are also often-used acceptance attributes. More than 90% of the respondents indicated they use the same test methods for QC and acceptance and more than half use the same point of sampling. Almost 90% of the respondents use pay adjustment procedures. Twenty-three of those using pay adjustments tend to use stepped pay schedules and 32 use both an incentive and disincentive. Sixty percent of the agencies use PWL (or PD) as the quality measure. Of the 29 agencies that use contractor tests in the acceptance decision, 11 compare one contractor test with one agency test, 7 use the *F*- and *t*-tests, and 14 use one agency test compared with several contractor tests. For verification, nine use independent samples, nine use split samples, and seven use both.

- Forty percent of the agencies use QA programs for portland cement concrete paving (PCCP), in which the contractor controls the quality and the agency performs acceptance. The other 60% are evenly divided between the practice of the agency controlling quality and performing acceptance and the contractor controlling the quality and the agency using contractor test results in the acceptance decision. Fifteen use materials and methods-type provisions. Gradation, air content, and slump are the most frequently used QC attributes. Sixty-five percent of the agencies establish the frequency for contractors to conduct QC tests and seven agencies require control charts. Thickness, air content, cylinder strength, slump, and gradation are the most often used attributes for acceptance. Beam strength, water–cement ratio, and ride quality are also often-used acceptance attributes. Eighty percent use the same test methods for QC and acceptance, and 60% use the same point of acceptance. Seventy percent of the respondents use pay adjustment procedures. Twenty-one of those using pay adjustments tend to use stepped pay schedules, 16 use both an incentive and disincentive, and 12 use only a disincentive. Appreciably fewer agencies use PWL (or PD) as the quality measure for PCCP than for HMA. Of the 14 agencies that use contractor tests in the acceptance decision, 6 compare one contractor test with one agency test, 3 use the *F*- and *t*-tests, and 5 use one agency test compared with several contractor tests. For verification, five use independent samples, four use split samples, and four use both. When training is required, most agencies use in-house training and certification for both agency and contractor personnel.
- The questionnaire results indicate substantial agreement between the types of QA programs used for PCC structures and PCCP. As for PCCP, almost 40% of the agencies use QA programs for PCC structures in which the contractor controls the quality and the agency performs acceptance. The other 60% are evenly divided between the practice of the agency controlling quality and performing acceptance and the contractor controlling the quality and the agency using contractor test results in the acceptance decision. Twenty-five agencies use materials and methods-type provisions. Gradation, air content, and slump are the most frequently used QC attributes. Almost 70% of the agencies establish the frequency for contractors to conduct QC tests and five agencies require control charts. Air content, cylinder strength, slump, and gradation are the most often used attributes for acceptance. More than 80% use the same test methods for QC and acceptance and 60% use the same point of sampling. More than 60% of the respondents use pay adjustment procedures. Fourteen of those using pay adjustments use stepped pay schedules, with 9 using both an incentive and disincentive and 19 using only a disincentive. The average, range, and PWL/PD are the quality measures most often used. Of the 14 agencies that use contractor tests in the acceptance decision, 7 compare one contractor test result with one agency test result, one uses the *F*- and *t*-tests, and 10 use one agency test result compared with several contractor test results. For verification, two use independent samples, four use split samples, and seven use both.
- The implementation of the independent assurance (IA) function is as diverse among agencies as other aspects of the QA program. This is evidenced by both the total full-time equivalent staffing levels that vary from 4 to 35 and from 0.5 to 16 as the total full-time equivalent per hundred million dollars of the construction and maintenance budget.
- The use of consultants is widespread, with more than 75% of the responding agencies reporting that they use consultants. This is not surprising considering the general downsizing that has taken place within state highway agencies (SHAs). Most use the consultants in place of or as a supplement to agency acceptance testing, and a significant number use them in place of or as a supplement to contractor QC testing.

  The use of innovative practices is not widespread. Eight agencies use warranties on a routine basis, and seven customarily use design–build. No other innovative practices were used to an appreciable extent.
- Twenty-three agencies reported that they anticipate significant changes in their QA programs in the near future and 22 indicated that they did not anticipate any changes. The products where changes are expected vary from entire pavement QA programs to individual material/construction components.

As discussed previously, QA programs are diverse. So diverse that many aspects of such programs could not be captured in this synthesis. Therefore, suggestions for future topics in education, research, and possibly an additional synthesis in the QA area are included here.

- Although the general understanding of QA programs has improved over the last decade, there is still room for improvement. Over the last several decades, QA educational programs have been available and many SHAs have taken advantage of them. Thus, it appears that some of this lack of understanding can be attributed to the turnover of personnel that all SHAs have undergone in the last decade. The terms used relative to QA are not commonplace terms used in most material/construction areas; therefore, they are sometimes interpreted differently or often misinterpreted. Thus, there is a continuing need to encourage the use of QA terms in the proper context.
- The large number of agencies using the same test method and point of sampling for QC and acceptance raises the question, "Why are these tests being run in this manner?" If it is simply to have two estimates of an attribute by two different parties to the contract it would appear that sampling and testing funds could be better used in other ways. Thus, there is a continuing need for training to use QA tests in a manner that provides optimum information and cost–benefit ratio.
- Questions concerning risks were not addressed in the questionnaire for this synthesis. Therefore, any conclusion as to the status of risk analysis within SHAs is somewhat speculative. The risks inherent in accepting results of contractor tests used in the acceptance decision without adequate verification does not appear to be well understood based on the procedures being used by many agencies. There is a need to demonstrate these risks in a user-friendly mode. One way of doing this is to improve computer programs so that agencies can determine the risks for the system they are using.
- As discussed throughout this synthesis, QA programs are complex. Attempting to cover all programs in a single synthesis would have required an unmanageable questionnaire. However, review of some areas related to QA programs that were not included in the questionnaire in this synthesis could be enlightening, including more in-depth knowledge of pay adjustment schedules, information on dispute resolution practices, and the manner in which control charts are used and control limits established.

REFERENCES

1. Final Rule, 23 CFR 637, "Quality Assurance Procedures for Construction," *Federal Register*, Vol. 60, No. 125, June 29, 1995, p. 33712.
2. Sabol, S.A. and R.L. Nickerson, *Research Results Digest 274: Quality Assurance of Structural Materials*, Transportation Research Board, National Research Council, Washington, D.C., Aug. 2003, 21 pp.
3. Bowery, F.E., Jr., and S.B. Hudson, *NCHRP Synthesis of Highway Practice 38: Statistically Oriented End-Result Specifications*, Transportation Research Board, National Research Council, Washington, D.C., 1976, 40 pp.
4. Hughes, C.S., R. Cominsky, and E.L. Dukatz, "Quality Initiatives," *Asphalt Paving Technology 1999*, Journal of the Association of Asphalt Paving Technologists, pp. 278–297.
5. Kopac, P.A., "Making Roads Better and Better," *Public Roads*, July/Aug. 2002, Vol. 66, No. 1, pp. 25–29.
6. Statistically Oriented End-Result Specifications, Associated General Contractors Position, Aug. 1977 (addendum to Reference 2).
7. Halstead, W.J., *NCHRP Synthesis of Highway Practice 65: Quality Assurance*, Transportation Research Board, National Research Council, Washington, D.C., 1979, 42 pp.
8. *AASHTO Quality Assurance Guide Specification*, AASHTO Quality Construction Task Force, American Association of State Highway and Transportation Officials, Washington, D.C., Feb. 1996, 29 pp.
9. Smith, G.R., *NCHRP Synthesis of Highway Practice 263: State DOT Management Techniques for Materials and Construction Acceptance*, Transportation Research Board, National Research Council, Washington, D.C., 1998, 51 pp.
10. Chamberlin, W.P., *NCHRP Synthesis of Highway Practice 212: Performance-Related Specifications for Highway Construction and Rehabilitation*, Transportation Research Board, National Research Council, Washington, D.C., 1995, 48 pp.
11. Kopac, P.A. and T.P. Teng, "Status of Performance-Related Specifications in the US," *Routes/Roads*, No. 315, *World Road Association–PIARC*, France, July 2002, pp. 57–67.
12. Burati, J.L., R.M. Weed, C.S. Hughes, and H.S. Hill, *Optimal Procedures for Quality Assurance Specifications*, Office of Research, Development, and Technology, Report FHWA-RD-02-095, Federal Highway Administration, McLean, Va., June 2003, 347 pp.
13. *Transportation Circular Number E-C037: Glossary of Quality Assurance Terms*, Transportation Research Board, National Research Council, Washington, D.C., Apr. 2002, 29 pp.
14. Miller–Warden Associates, *A Plan for Expediting the Use of Statistical Concepts in Highway Acceptance Specifications*, Office of Research and Development, Bureau of Public Roads, U.S. Department of Commerce, Washington, D.C., Aug. 1963.
15. *The Statistical Approach to Quality Control in Highway Construction*, U.S. Department of Commerce, Bureau of Public Roads, Washington, D.C., Apr. 1965, 43 pp.
16. Miller–Warden Associates, *NCHRP Report 17: Development of Guidelines for Practical and Realistic Construction Specifications*, Transportation Research Board, National Research Council, Washington, D.C., 1965, 109 pp.
17. Burati, J.L., Jr., and C.S. Hughes, *Highway Materials Engineering, Materials Control and Acceptance—Quality Assurance*, Report FHWA-HI-90-004, National Highway Institute, McLean, Va., Feb. 1990, 170 pp.
18. Dunbar, D.W., "Legal Responsibilities for Quality Control in Highway Construction," *Western Builder*, Aug. 1966, pp. 51–57.
19. *AASHTO Implementation Manual for Quality Assurance*, AASHTO Quality Construction Task Force, American Association of State Highway and Transportation Officials, Washington, D.C., Feb. 1996, 84 pp.
20. *Acceptance Sampling Plans for Highway Construction*, Draft Standard Recommended Practice R-9-04, American Association of State Highway and Transportation Officials, Washington, D.C., 2004, 73 pp.
21. *Standard Recommended Practice for Technician Training and Qualification Programs*, AASHTO Materials, Part I Specifications: R-25-00, American Association of State Highway and Transportation Officials, Washington, D.C., 2000.
22. *Program Detail Manual for Certification in the Field of Transportation Engineering Technology, Subfield of Highway Materials*, 5th ed., National Institute for Certification in Engineering Technologies, Alexandria, Va., 1996.
23. *Program Detail Manual for Certification in the Field of Construction Materials Testing, Subfield of Asphalt, Concrete, and Soils*, 6th ed., National Institute for Certification in Engineering Technologies, Alexandria, Va., 1997.
24. *Guidelines for Establishing a Technician Training and Certification Program*, National Quality Initiative, Toronto, ON, Canada, Sep. 1997.
25. "Pavement Technology," Federal Highway Administration, Washington, D.C., July 1998 [Online]. Available: http://www.fhwa.dot.gov/pavement/techqual.htm. [Nov. 15, 2004].
26. Federal Highway Administration, Washington, D.C. [Online]. Available: http://www.fhwa.dot.gov/pavement/labqual.htm.

27. *Against All Odds; Inside Statistics,* Television and Video Course, Annenburg/CPB Collection, Washington, D.C., 1989.
28. *ASTM Manual on Presentation of Data and Control Chart Analysis*, 6th ed., Series MNL 7, Committee E-11 on Quality and Statistics, American Society for Testing and Materials, West Conshohocken, Pa., 106 pp.
29. Ernzen, J. and K. Vogelsang, "Contractor-Led Quality Control and Quality Assurance Plus Design-Build: Who Is Watching the Quality?" *Transportation Research Record 1813*, Transportation Research Board, National Research Council, Washington, D.C., 2002, pp. 253–259.
30. Afferton, K.C., J. Freidenrich, and R. Weed, "Managing Quality: Time for a National Policy," *Transportation Research Record 1340*, Transportation Research Board, National Research Council, Washington, D.C., 1992, pp. 3–39.
31. *Sampling Procedures and Tables for Inspection by Variables for Percent Defective*, Military Standard 414, U.S. Department of Defense, Washington, D.C., 1957.
32. Burati, J.L., W.C. Bridges, and S.A. Ackerman, "Evaluation of Quality Assurance Programs for Bituminous Paving Mixtures," Final Project Report, FHWA-SC-95-02, Clemson University, Clemson, S.C., May 1995, 110 pp.
33. Byrne, M.P., J.H. Wang, and D.M. Shao, "Evaluation of Independent Assurance Sampling and Testing Variation Limits," Final Report, FHWA-RIDOT-RTD-03-5, University of Rhode Island, Kingston, Apr. 2003, 24 pp.
34. Wegman, D.E., "Minnesota's Quality Management Program: A Process for Continuous Improvement," Quality Management of Hot Mix Asphalt, *ASTM STP 1299*, D.S. Decker, Ed., American Society for Testing and Materials, West Conshohocken, Pa., 1996, pp. 1–3.
35. Anderson, S.D. and J.S. Russell, *NCHRP Report 451: Guidelines for Warranty, Multi-Parameter, and Best Value Contracting*, Transportation Research Board, National Research Council, Washington, D.C., 2001, 101 pp.
36. Scott, S., "Contract Management Techniques for Improving Construction Quality," Final Report, FHWA-RD-97-067, Office of Engineering Research & Development, Federal Highway Administration, McLean, Va., July 1997, 237 pp.
37. *Transportation Research Circular 386: Innovative Contracting Practices*, Transportation Research Board, National Research Council, Washington, D.C., Dec. 1991, 74 pp.
38. Ernzen, J. and K. Vogelsang, "Evaluating Design-Build Procurement Documents for Highway Projects: How Good Are They?" *Transportation Research Record 1761,* Transportation Research Board, National Research Council, Washington, D.C., 2001, pp. 148–158.
39. McMahon, T.F., W.J. Halstead, W.W. Baker, E.C. Granley, and J.A. Kelly, "Quality Assurance in Highway Construction," initially compiled in Feb./Dec. 1969, *Public Roads*, Vol. 35, Nos. 6–11, reprinted in FHWA-TS-89-038, Oct. 1990, 60 pp.
40. Sherman, G.B., R.O. Watkins, and R. Prysock, *A Statistical Analysis of Embankment Compaction*, California Department of Public Works, Division of Highways, Sacramento, May 1966, 23 pp.
41. Wolters, R.O., *Modern Concepts for Density Control, Phase III: Embankment Materials*, Final Report, MN HW 5-180 74-3, Minnesota Department of Highways, St. Paul, 1973, 51 pp.
42. Zahur, S., D.A. Anderson, and C.E. Antel, "Development of a Statistically Based Specification for Unbound Aggregates," *Transportation Research Record 1056*, Transportation Research Board, National Research Council, Washington, D.C., 1986, pp. 1–9.
43. Anday, M.C., "Statistical Specification for Acceptance of Pug Mill Mixed and Subbase Materials," *Highway Research Record 357*, Highway Research Board, National Research Council, Washington, D.C., 1971, pp. 39–52.
44. Dillard, J.H., *Field Experiment with Statistical Methods,* Virginia Highway Research Council, Charlottesville, 1986, pp. 1–9.
45. Afferton, K., "A Statistical Study of Asphaltic-Concrete," *Highway Research Record 184*, Highway Research Board, National Research Council, Washington, D.C., 1967, pp. 13–24.
46. Mills, W.H. and O.S. Fletcher, "A System for Control and Acceptance of Bituminous Mixtures by Statistical Methods," *Highway Research Record 184*, Highway Research Board, National Research Council, Washington, D.C., 1967, pp. 24–54.
47. Hughes, C.S., *NCHRP Synthesis of Highway Practice 232: Variability in Highway Pavement Construction,* Transportation Research Board, National Research Council, Washington,w D.C., 1996, 38 pp.
48. *Determination of Statistical Parameters for Highway Construction,* Research Project No. 18, State Road Commission of West Virginia, by Materials, Research and Development, Inc., Miller–Warden Associates Division, Raleigh, N.C., July 1966, 64 pp.
49. Adam, V. and S.C. Shah, *Tolerances for Asphaltic Concrete and Base Courses*, Louisiana Department of Highways, Baton Rouge, Nov. 1966, 27 pp.
50. Solaimanian, M., W.E. Elmore, and T.W. Kennedy, "Effectiveness Comparison of TxDOT Quality Control/Quality Assurance and Method Specifications," Office of Research and Technology, Texas Department of Transportation, Austin, Dec. 1998, 116 pp.
51. Ksaibati, K. and N. Butts, "Evaluating the Impact of QC/QA Programs on the Asphalt Mixture Variability," MPC Report No. 03-146, University of Wyoming, Laramie, June 2003, 69 pp.
52. Benson, P.E., "Comparison of End Result and Method Specifications for Managing Quality," *Transportation Research Record 1491*, Transportation Research Board, National Research Council, Washington, D.C., 1995, pp. 3–10.

53. Uhlmeyer, J.S., "PCCP Intersections Design and Construction in Washington State," Final Report, WA-RD 503.1, Washington State Transportation Center, University of Washington, Seattle, May 2001, 419 pp.
54. Graveen, C., J. Weiss, J. Olek, T.E. Nantung, and V.L. Gallivan, "Implementation of Performance-Related Specification for Concrete Pavement in Indiana," Presented at the 83rd Annual Meeting of the Transportation Research Board, Washington, D.C., Jan. 2004, 27 pp.
55. Wilde, W.J., B.F. McCullough, and D.W. Fowler, "Development of a Quality Assurance Program for Portland Cement Concrete in South Carolina," FHWA-SC-01-01, Final Report, South Carolina Department of Transportation, Jan. 2001, 69 pp.
56. FHWA Technical Advisories, Federal Highway Administration, Washington, D.C. [Online]. Available: http://www.fhwa.dot.gov/legsregs/directives/techadvs.htm [Dec. 8, 2004].
57. Benson, P.E., "Process for Selecting Innovative Materials Quality Assurance Practices," *Transportation Research Record 1900*, Transportation Research Board, National Research Council, Washington, D.C., 2004, pp. 67–78.

APPENDIX A
23 CFR 637

Federal Register: June 29, 1995 (Volume 60, Number 125, p. 33712)

Section: Rules and Regulations
Agency: **Federal Highway Administration (FHWA), DOT**.
Title: QUALITY ASSURANCE PROCEDURES FOR CONSTRUCTION
Action: Final rule.

DEPARTMENT OF TRANSPORTATION
Federal Highway Administration
23 CFR Part 637
[FHWA Docket No. 9413]
RIN 2125AD35

Quality Assurance Procedures for Construction

SUMMARY: The FHWA is revising its regulations that establish general requirements for quality assurance procedures for construction on Federal-aid highway projects. The rule provides more flexibility than the existing regulation. The rule allows the use of contractor test results in making the acceptance decision and allows the use of consultants in the independent assurance program and verification sampling and testing. The regulation requires testers and laboratories to be qualified. However, it gives the States the flexibility to establish those qualifications. The revisions will clarify existing policy and procedures and provide additional guidance on the use of contractor-supplied test results in acceptance plans.

EFFECTIVE DATE: July 31, 1995.

FOR FURTHER INFORMATION CONTACT: Mr. Michael Rafalowski, Office of Engineering, HNG23, 202-366-1571; or Mr. Wilbert Baccus, Office of the Chief Counsel, HCC32, 202-366-0780; Federal Highway Administration, 400 Seventh Street, SW, Washington, DC 20590. Office hours are 7:45 a.m. to 4:15 p.m., e.t., Monday through Friday, except Federal holidays.

SUPPLEMENTARY INFORMATION:

BACKGROUND

The current regulations on sampling and testing of materials and construction appear in 23 CFR Part 637, Construction Inspection and Approval. These regulations were last revised in January 1987. The regulations were written using the concept of the State performing all the sampling and testing, which had been the traditional approach to sampling and testing. The regulations do not address the use of contractor testing. As a result, a number of questions arose in those States which were using contractor testing in their quality control/quality assurance (QC/QA) programs.

The existing regulations do not recognize the use of contractor testing results in an acceptance program. An acceptance program is the process of determining whether the materials and workmanship are in reasonably close conformity with the requirements of the approved plans and specifications. In 1992, the FHWA studied the ramifications of using contractor-performed sampling and testing results. The results of its study are reported in "Limits of Use of Contractor Performed Sampling and Testing," dated July 1, 1993. (A copy of the report is available in the docket for inspection and copying.) One of the report's recommendations was that contractor sampling and testing may be used in acceptance programs, provided adequate checks and balances are in place to protect the public investment. The revisions to Part 637 made in this final rule would implement the committee's recommendation.

This final rule provides more flexibility to the States in designing their acceptance programs than currently exists. Acceptance of materials and construction will not be based solely on any one set of information. Each State's verification sampling and testing will be used to ensure the quality of the product. In addition, the rule will permit the use of data from the contractors' quality control sampling and testing programs in acceptance programs if the results from the States' verification sampling and testing programs confirm the quality of the material. The verification sampling and testing must be performed on independent samples obtained by the State or designated agent to verify the quality of the material. If the results of a State's verification sampling and testing program do not confirm the quality of the product, a dispute resolution system must be used to determine payment to the contractor.

The requirement for an independent assurance (IA) program will remain in place. The rule will provide the States more flexibility in designing their IA program. The IA program will allow the use of witnessing, split samples, proficiency samples, and equipment calibration as an independent check of the field sampling and testing procedures and equipment to assure that the testing is being performed properly by both the State and the contractor personnel.

COMMENTS TO THE DOCKET

A notice of proposed rulemaking (NPRM) was published in the Federal Register on July 12, 1994 (59 FR 35493), in which the FHWA proposed to revise 23 CFR Part 637, Construction Inspection and Approval. A total of 50 commenters responded to the NPRM as follows: 35 State highway agencies, 1 local agency, 1 toll authority, 10 construction industry associations and contractors, and 3 Subcommittees of the American Association of State Highway and Transportation Officials (AASHTO). The major comments and the FHWA's response thereto are summarized as follows.

Supportive of Change

Twenty-six commenters expressed their support for the revisions to the regulation. Fifteen commenters provided comments without indicating support or opposition to the NPRM. The remaining nine commenters were generally opposed to the proposed rule.

Use of Contractor Test Results

Commenters expressed three related concerns over the required system of checks and balances employed when contractor test results are used in the acceptance decision: 1) requiring the use of independent samples instead of allowing either independent samples or split samples; 2) requiring the use of the F-test and the t-test (which are standard statistical tests for comparing the variances and means of two sets of data) because of the complexity of using the statistical tests; and 3) the perceived duplication of effort between the verification sampling and testing and the testing required by covering the contractor sampling and testing program in the IA program.

The overall intent of the program is to provide adequate assurance that the public is receiving the desired quality in the product produced by the contractor. The first level of assurance is provided by qualifying laboratories and testing personnel. This assures that the equipment and personnel are capable of performing the tests properly. The second level of assurance is provided by the IA program. This level assures that the testers and equipment remain capable of performing the tests properly. The third level of assurance is provided by verification sampling and testing. This level assures the quality of the product.

There appears to have been some misunderstanding of the total level of effort required. The rule as adopted gives the States wide latitude in designing the acceptance program. The system approach to IA assures the capabilities of all equipment and testers regardless of the number of projects or material quantities involved. A broad interpretation of the existing regulations would allow the system approach to IA. However, the final rule explicitly allows the system approach to IA. In those States that are performing a significant amount of testing on split samples and no testing on independent samples, testing on split samples would remain as IA sampling and testing; however, some verification testing on independent samples would be required to confirm the quality of the product. In addition, the verification of the quality of the material can be performed on a mix design or grading of material from a given source and is not limited to project-specific data.

Eleven commenters expressed concern over requiring the use of independent samples for the verification sampling and testing program. The commenters recommended that the use of split samples be permitted for the verification sampling and testing program. The commenters are concerned about the potential problems that may arise with differences in testing results caused by sampling errors.

There are three sources of differences between two test results, differences in the material, differences in test procedures, and differences in sampling procedures. Split samples will only address the differences in test procedures and will only provide assurance that the contractor is performing the tests properly. In a balanced system it is also necessary to assure that sampling of materials is performed properly. It is our intent that the verification sampling and testing program be used to independently validate the quality of the material. Using independent samples will insure that all sources of differences are measured. The FHWA recognizes the need to ensure that each contractor performs the tests correctly; that is the reason for extending laboratory and testing personnel qualification requirements and IA program requirements to the contractor if the contractor's test results are to be used in the acceptance decision. The FHWA expects the testing variability between the contractor and the State to be held to a minimum by requiring the contractor's testing program to be covered by an IA program and requiring the testing personnel and laboratories to be qualified. The FHWA has changed the definition of "verification sampling and testing" and Section 637.207 (a)(1)(ii)(B) to clarify the fact that the verification sampling and testing program is being used to validate the quality of the material.

Eight commenters objected to requiring the use of the *F*-test and *t*-test for verifying a contractor's test data. The commenters were concerned about the complexity of the *F*-test and *t*-test, which would have to be used by field personnel and the lack of flexibility in allowing other comparison systems. The commenters requested that the regulation be revised to allow other types of comparison systems. The FHWA agrees with the concerns and has removed the requirement for a specific comparison procedure. Each State will have the latitude to develop its own verification system.

Three commenters—two State Highway Agencies (SHAs) and one local highway agency—objected to including contractors' testers in States' IA programs. The commenters are concerned over the additional resources involved in extending the IA program to contractor testing.

If a contractor's test results are to be used in the acceptance decision, assurance must be provided that the contractor's testers and equipment remain capable of performing the tests properly. Some States are currently performing split sampling and testing on project sites to validate the contractor's test results. This split sampling and testing would meet the requirements for an IA program on contractor testing. This proposed requirement has been retained in the final rule.

Qualified Sampling and Testing Personnel

Four commenters specifically supported the concept of certifying testing personnel.

Two commenters wanted to change the term certified personnel to qualified personnel. The FHWA agrees with the comments since the goal of the FHWA is to have qualified personnel perform the testing. The term "certified" was deleted from the definition of qualified testing personnel.

Sixteen commenters expressed concern about the cost, specific requirements, and/or two-year implementation period for establishing qualification programs for testing personnel. To allow adequate time to develop qualification programs, we have extended the implementation time from two years to five years. If a State chooses to use a certification program as its qualification program, the FHWA is developing training material that can be modified for State use. The FHWA will also assist the States in adapting the material for their use.

Independent Assurance Program

Thirteen commenters objected to the proposal to remove the requirement that SHA personnel perform IA testing. The States wanted to continue to perform IA testing as a means to maintain expertise in the materials sampling and testing area and maintain the credibility of their materials programs. Since materials sampling and testing are an essential part of determining the quality of the product that is obtained from the use of Federal-aid funds, the FHWA has an interest in maintaining the States' expertise and credibility. However, in cases where States are using contractor test results in acceptance decisions, the FHWA believes it is important that the States have the option of using consultants to perform IA testing. It is important to note that the final rule does not require a SHA to use consultants in the IA program, but simply gives SHAs the option to do so. The FHWA has added Section 637.205(b) which requires States to maintain an adequate, qualified staff with the capability of overseeing the entire quality assurance program and specifically requires the States to maintain a central laboratory. This requirement is consistent with 23 U.S.C. 302, which requires each State to maintain an adequate highway department.

Three commenters requested further clarification on the use of the system approach in performing an IA program. The intent of the system approach to the IA program is to concentrate on assuring that the testing personnel and equipment remain capable of performing the tests properly, regardless of the location or number of projects covered by the equipment and tester. The system approach will permit an SHA to fulfill the requirement for an IA program by implementing a schedule of activities to cover equipment operations and tester competence. The activities may include calibration checks, split samples, proficiency samples, and observations. The schedules and type of activity would be based on the test procedure. In the system approach, the frequency of IA may be independent of the number of tests performed or the quantity of material tested. It is envisioned that the system approach will be especially useful in cases where one tester performs testing for more than one project during a construction season. The previous requirement for IA entailed sampling and testing frequencies based on individual project production. In addition, a State may choose to use the information developed from the IA program in the qualification programs for testers and laboratories. One commenter asked if the NPRM would allow a State to use a hybrid approach, which would include some frequencies based on project quantities and frequencies based on the overall system. This rule as written would allow that approach. It should be noted that the rule does not require a State to use this approach.

One commenter wanted the requirements for the IA program to be less stringent. The requirements in the final rule for IA have been made less prescriptive than the current regulations and give a State more latitude in designing its IA system. The existing regulation requires State personnel to perform the IA sampling and testing. The final rule would allow: (1) the use of accredited consultant laboratories in executing an IA program, (2) a system approach instead of a project approach, (3) proficiency samples instead of split samples, and (4) equipment calibration to cover the testing equipment.

Laboratory Qualification

Four commenters supported the proposed requirements for laboratory qualifications.

Eight commenters expressed concerns about the requirements for laboratory qualifications. The NPRM proposed to include by reference two paragraphs from the "Standard Recommended Practice for Establishing and Implementing a Quality System for Construction Testing Laboratories" (R-18) published by the AASHTO in the "Standard Specifications for Transportation Materials and Methods of Sampling and Testing." The commenters believed that R-18 was not appropriate for field laboratories. It was not the FHWA's intent that the entire R-18 standard be used for the qualification of field laboratories. Due to the confusion caused by specifying only a part of R-18, the rule has been revised to specifically list the minimum requirements for field laboratories and delete the reference to R-18.

Eight commenters wanted clarification of the requirements for accreditation of the SHA central laboratory. It is the intent of the FHWA that the accreditation program must meet the guidelines in ASTM E994. In addition to the guidelines in ASTM E994, we have two additional concerns: First, regarding the acceptability of the assessors; and second, concerning the scope of the onsite assessment. For an accreditation program to be acceptable to the FHWA, the assessor must be employees of the accrediting body and not employed by a laboratory which may compete for work with the laboratory being assessed. This would avoid any potential conflicts of interest. In addition, the onsite assessment must include a detailed review of the test procedures in which the laboratory is being accredited. The FHWA believes that only one laboratory accreditation program currently meets the above concerns, and that is the AASHTO Accreditation Program. As we understand the operating procedures of other accreditation programs, they allow reviewers to be employees of other testing laboratories and do not require the laboratory to demonstrate all the tests in which the laboratory is being accredited. If other accreditation programs can satisfy our concerns, we will approve them. Any inquires or requests for approval should be directed to the FHWA's Office of Engineering.

Six commenters expressed concern about the cost and implementation time necessary for accrediting an SHA central laboratory. The commenters believe that two years is too short a time in which to become accredited. At this time 30 SHAs are accredited by the AASHTO Accreditation Program (AAP). The FHWA contacted the AAP to obtain data on the average length of time required by the AAP to accredit a SHA laboratory after receipt of an application for accreditation. Based on the information supplied by AAP, the FHWA believes that two years is an adequate lead time for obtaining accreditation. The requirement for accreditation replaces the inspections by the National Reference Laboratories, which are required by Section 637.205 of the current regulation. The actual cost of accreditation to the SHA is the same as the cost of inspection program that it replaces. However, there will be some costs associated with developing the quality system for the initial accreditation for the SHAs. The rule provides flexibility to the SHAs to designate private laboratories to perform independent assurance tests and dispute resolution testing. Since the SHAs must review the qualifications of designated laboratories, the SHAs need to be qualified at the highest level, which is accreditation. Therefore, this final rule maintains the laboratory accreditation requirements as originally proposed.

Definitions

Four commenters suggested changes to the definition of quality control. The definition of quality control was adapted from the definition in ANSI 90 and ISO 9000 which are the industry consensus standards for quality assurance. Therefore, the FHWA is retaining the definition as proposed.

Two commenters wanted to delete the word "accredited" from the definition of "qualified laboratories." There appears to be confusion over the use of the term "accreditation" since the NPRM used the word to describe two different levels of qualifications. The FHWA agrees with the comment because of the apparent confusion. The word "accredited" has been removed from the definition of "qualified laboratories."

Two commenters wanted clarification of the term "vendor." A definition of "vendor" has been added to insure that it includes suppliers of project-produced materials. It was the FHWA's intent that the rule cover only project-produced materials and not manufactured materials.

One commenter suggested changes to the definition of "quality assurance." The definition of "quality assurance" was adapted from the definitions in the ANSI 90 and ISO 9000 standards, which are the industry consensus standards for quality assurance. Therefore, the FHWA has retained this definition as proposed in the NPRM.

One commenter suggested requiring random sampling. The FHWA agrees with the comment. In order for test data used in the acceptance decision to be properly analyzed, samples must be obtained on a random basis. Section 637.205(e) has been added to require random sampling.

One commenter was concerned with the wording of the definition for IA, which the commenter interpreted as requiring the IA to be performed by a consultant. As stated earlier, it is the FHWA's intent that the States have the option to perform IA sampling and testing themselves or have a qualified designated agent perform the testing. The definition in the final rule has been revised to reflect our intent.

Miscellany

Eight commenters requested a delay in issuing a final rule. Their major concern was over potential conflicts between this final rule and AASHTO's effort to develop guide specifications for quality assurance. The AASHTO effort is related to this rulemaking. However, the "AASHTO Quality Assurance Guide Specification" and the "AASHTO Implementation Manual for Quality Assurance" are in the draft stage and are still being reviewed. It may be some time before these documents receive full endorsement by AASHTO. Since the current regulations do not address the practice of using contractor testing in making acceptance decisions, the FHWA believes that it is necessary to proceed with the final rule. The commenters were also concerned that the SHAs did not have adequate time to comment on the regulation. The NPRM provided a 60 day comment period. All comments that were received by the FHWA, including the eleven received after the closing of the comment period, were considered and included in the analysis. In addition, the FHWA received comments from 35 of the 52 SHAs. Therefore, the FHWA believes that adequate time was provided.

Five commenters provided comments on the dispute resolution system. There were comments on both sides of the issue of whether the dispute resolution system should allow third party involvement. Three commenters were in favor of keeping the system in the State; two were in favor of using third parties. In the NPRM the FHWA proposed to permit the SHAs to determine how they wanted to set up the dispute resolution system. The FHWA is aware of cases where a dispute resolution system has worked well in both cases, so this proposal has been retained in the final rule.

Three commenters requested clarification of the terms "acceptance," "verification," and "assurance." This rule requires an acceptance program, which includes the establishment of qualifications of testers and laboratories and inspection of construction operations and testing performed by the SHA or its designated agent. Verification sampling and testing is used to validate the quality of the product. Independent assurance is used specifically to insure that the testing is performed correctly and that the equipment is in calibration.

Two commenters provided comments on the materials certificate. One commenter requested that the wording on the material certificate be revised from requiring the materials and operations to be in "conformity with the approved plans and specifications" to "reasonably close conformity to the approved plans and specification." The commenter was concerned about the added work of adding the individual material exceptions to the project plans and specifications to the materials certificate. The current regulation requires the material certificate to list all materials that do not meet the specifications. The FHWA reserves the right to review the materials certificate to determine if the materials are in conformity with the project plans and specifications. Therefore, the FHWA has retained the wording as proposed in the NPRM. The other commenter wanted to eliminate the requirement for the materials certificate. Section 637.201 limits the rule to projects on the NHS. In addition, Section 637.207(a)(3) further limits the requirement for a materials certificate to projects that are subject to FHWA oversight reviews. This will eliminate the requirement for a materials certificate for the vast majority of projects. Since the cost of materials make up a substantial portion of each project and the information supplied by the materials certificate indicates the quality of the material, it is necessary to have the materials certificate in order to make an informed decision on whether to accept those projects for which the FHWA has retained construction oversight. Therefore, the FHWA has retained the proposed requirement for a materials certificate in this final rule.

One commenter indicated that the cost of implementing the regulation was high and a full regulatory review was needed. As noted below the FHWA has determined that this action is not a significant regulatory action under Executive Order 12366, Regulatory Planning and Review, nor significant under DOT Order 2100.5, Policies and Procedures for Simplification, Analysis, and Review of Regulations, and has concluded that a full regulatory evaluation is not required.

Costs to the States. Currently all States must have approved sampling and testing programs which include an IA program. In addition, all States are required to have their central laboratories inspected by the National Reference Laboratories. As indicated in the fee schedule for the AAP, the actual cost of accreditation itself for the SHAs is the same as the current inspection fees. The additional cost to the States for becoming accredited is in developing the quality assurance manuals, which are required by the AAP. The justification for requiring accreditation is stated above. Since the vast majority of States have qualification requirements for their subsidiary laboratories, there would be no additional costs for the States that have these requirements. There would be minimal costs to those States that will have to develop qualification requirements for laboratories. There would be some costs in developing qualifications for testers. One aspect of tester qualifications is attendance at training programs. All States have some training for their technicians, but some of this training may have to be upgraded. However, as stated earlier, the FHWA has a training effort that is available to assist the States in setting up certification programs. The certification programs could be used in the States' establishment of tester qualifications.

Costs to the public. There would be no additional costs to the industry if a State chooses not to incorporate contractor tests into the acceptance system. If a State chooses to use contractor tests in acceptance decisions, contractors would be required to hire employees qualified in the appropriate tests and the State would be required to ensure that the contractors maintain a qualified laboratory or hire a qualified laboratory to perform the testing. When a State uses contractor quality control testing results in the acceptance decision, testing performed by the State is reduced. This reduction in testing by the State reduces the overhead costs in the State. However, any additional cost the contractors incur in performing the testing, including costs of obtaining qualified laboratories and testers, will be passed onto the State through higher bid prices. The cost savings by the State due to the reduction of testing by State personnel would be offset by the increase in bid prices charged by the contractors. As a result, the FHWA believes that the additional costs of these actions would be minimal.

One commenter was concerned because its quality assurance program is located in several documents and it did not want to consolidate the information into one document. The FHWA does not see the need for all the documentation of a State's quality assurance program to be in one document.

One commenter interpreted the NPRM to propose a requirement for a central laboratory and the commenter opposed such a requirement. The NPRM did not expressly propose to require a central laboratory; however, the NPRM did propose to require that each State's central laboratory be accredited by the AAP or a comparable program approved by the FHWA. For the reasons stated above, this final rule now requires a central laboratory.

One commenter was concerned about the effect of these QC/QA regulations on small projects. As indicated in the preamble of the NPRM, it is not the intent of the FHWA in this regulation to require the use of contractor testing in the acceptance decision. In addition, the rule expressly covers only projects on the National Highway System (NHS); projects not on the NHS can use other SHA procedures to accept materials. It is anticipated that the majority of small projects will not be on the NHS.

One commenter was against QC/QA procedures. The rule does not require SHAs to use statistical concepts or to use contractor-supplied test results in the acceptance decision. However, the rule does establish minimum requirements if an SHA chooses to use contractor tests results in the acceptance decision.

One commenter suggested a revision to the portion of Section 637.207 concerning inspection to reflect the positive as well as the negative aspects of the quality of the product or construction. The section in the NPRM read, "The SHA shall inspect the product or construction or both for attributes that are detrimental to the performance of the finished product." The FHWA agrees with the comment. Section 637.207(a)(1)(i)(C) has been revised to reflect both beneficial and negative aspects of the quality of the finished product.

One commenter indicated that the regulation was too prescriptive. The rule, however, provides more flexibility than the existing regulation. The rule allows the use of contractor test results in making the acceptance decision and allows the use of consultants in the independent assurance program. Neither of these were allowed by the existing regulations. The regulation requires testers and laboratories to be qualified. However, it gives the States the flexibility to establish those qualifications. In addition, the final rule modified Section 637.207 to remove the requirement for a specific comparison procedure to validate the quality of the material. The rule clarifies existing policy and procedures and provides additional guidance on the use of contractor-supplied test results in acceptance plans.

One commenter questioned the title and purpose of the proposed rule, indicating that the rule covers materials and not construction. Over 50 percent of the cost of construction is the cost of the material. In addition, the rule requires each State to inspect construction to insure that the construction procedures do not adversely affect the properties of the material. Therefore, the title of this rule remains unchanged. Executive Order 12866 (Regulatory Planning and Review) and DOT Regulatory Polices and Procedures. The FHWA has determined that this action is not a significant regulatory action within the meaning of Executive Order 12866 or significant within the meaning of Department of Transportation's regulatory policies and procedures. The FHWA, at 23 CFR 637, currently has regulations covering sampling and testing. The rule provides the States with additional flexibility in comparison to the current regulations. States will be allowed to use contractor test results in making acceptance decisions and consultants to perform independent assurance testing. Other changes update the current regulations to accommodate contractor-performed sampling and testing and reinforce existing policy. Therefore, it is anticipated that the economic impact of this rulemaking will be minimal and a full regulatory evaluation is not required.

Regulatory Flexibility Act

In compliance with the Regulatory Flexibility Act (5 U.S.C. 601612), the FHWA has evaluated the effects of this action on small entities. The FHWA concluded that this action may provide some small testing firms with an opportunity to perform more work than was allowed by the previous regulations. Although the regulation will have a positive impact on these testing firms, the number of firms affected will be small and the amount of additional work would be insignificant. Therefore, the FHWA hereby certifies that this rulemaking will not have a significant economic impact on a substantial number of small entities. Executive Order 12612 (Federalism Assessment)

This action has been analyzed in accordance with the principles and criteria contained in Executive Order 12612. The rule provides the States with additional flexibility over the current regulations. States will be allowed to use contractor test results in making acceptance decisions and consultants to perform IA testing. Therefore, it has been determined that this action does not have sufficient federalism implications to warrant the preparation of a separate federalism assessment.

Executive Order 12372 (Intergovernmental Review)

Catalog of Federal Domestic Assistance Program Number 20.205, Highway Planning and Construction. The regulations implementing Executive Order 12372 regarding intergovernmental consultation on Federal programs and activities apply to this program.

Paperwork Reduction Act

This action does not contain a collection of information requirement for purposes of the Paperwork Reduction Act of 1980, 44 U.S.C. 3501-3520.

National Environmental Policy Act

This rulemaking does not have any effect on the environment. It does not constitute a major action having a significant effect on the environment, and therefore does not require the preparation of an environmental impact statement pursuant to the National Environmental Policy Act of 1969 (42 U.S.C. 4321 et seq.).

Regulation Identification Number

A regulation identification number (RIN) is assigned to each regulatory action listed in the Unified Agenda of Federal Regulations. The Regulatory Information Service Center publishes the Unified Agenda in April and October of each year. The RIN contained in the heading of this document can be used to cross reference this action with the Unified Agenda.

List of Subjects in 23 CFR Part 637

Grant programs—transportation, highways and roads, quality assurance, materials sampling and testing.
Issued on: June 22, 1995.

RODNEY E. SLATER,
Federal Highway Administrator

In consideration of the foregoing, the FHWA is amending title 23, Code of Federal Regulations, by revising part 637 to read as follows:

PART 637—CONSTRUCTION INSPECTION AND APPROVAL
Subpart A—[Reserved]
Subpart B—Quality Assurance Procedures for Construction
Sec.
637.201 Purpose.
637.203 Definitions.
637.205 Policy.
637.207 Quality assurance program.
637.209 Laboratory and sampling and testing personnel qualifications.
Appendix A to Subpart B—Guide Letter of Certification by State Engineer
Authority: 23 U.S.C. 109, 114, and 315; 49 CFR 1.48(b).
Subpart A—[Reserved]
Subpart B—Quality Assurance Procedures for Construction

Sec. 637.201 Purpose.

To prescribe policies, procedures, and guidelines to assure the quality of materials and construction in all Federal-aid highway projects on the National Highway System.

Sec. 637.203 Definitions.

Acceptance program. All factors that comprise the State highway agency's (SHA) determination of the quality of the product as specified in the contract requirements. These factors include verification sampling, testing, and inspection and may include results of quality control sampling and testing.

Independent assurance program. Activities that are an unbiased and independent evaluation of all the sampling and testing procedures used in the acceptance program. Test procedures used in the acceptance program, which are performed in the SHA's central laboratory would not be covered by an independent assurance program.

Proficiency samples. Homogeneous samples that are distributed and tested by two or more laboratories. The test results are compared to assure that the laboratories are obtaining the same results.

Qualified laboratories. Laboratories that are capable as defined by appropriate programs established by each SHA. As a minimum, the qualification program shall include provisions for checking test equipment and the laboratory shall keep records of calibration checks.

Qualified sampling and testing personnel. Personnel who are capable as defined by appropriate programs established by each SHA.

Quality assurance. All those planned and systematic actions necessary to provide confidence that a product or service will satisfy given requirements for quality.

Quality control. All contractor/vendor operational techniques and activities that are performed or conducted to fulfill the contract requirements.

Random sample. A sample drawn from a lot in which each increment in the lot has an equal probability of being chosen.

Vendor. A supplier of project-produced material that is not the contractor.

Verification sampling and testing. Sampling and testing performed to validate the quality of the product.

Sec. 637.205 Policy.

(a) Quality assurance program. Each SHA shall develop a quality assurance program, which will assure that the materials and workmanship incorporated into each Federal-aid highway construction project on the NHS are in conformity with the requirements of the approved plans and specifications, including approved changes. The program must meet the criteria in Sec. 637.207 and be approved by the FHWA.

(b) SHA capabilities. The SHA shall maintain an adequate, qualified staff to administer its quality assurance program. The State shall also maintain a central laboratory. The State's central laboratory shall meet the requirements in Sec. 637.209(a)(2).

(c) Independent assurance program. Independent assurance samples and tests or other procedures shall be performed by qualified sampling and testing personnel employed by the SHA or its designated agent.

(d) Verification sampling and testing. The verification sampling and testing are to be performed by qualified testing personnel employed by the SHA or its designated agent, excluding the contractor and vendor.

(e) Random samples. All samples used for quality control and verification sampling and testing shall be random samples.

Sec. 637.207 Quality assurance program.

(a) Each SHA's quality assurance program shall provide for an acceptance program and an independent assurance (IA) program consisting of the following:
(1) Acceptance program.
(i) Each SHA's acceptance program shall consist of the following:

(A) Frequency guide schedules for verification sampling and testing, which will give general guidance to personnel responsible for the program and allow adaptation to specific project conditions and needs.
(B) Identification of the specific location in the construction or production operation at which verification sampling and testing is to be accomplished.
(C) Identification of the specific attributes to be inspected which reflect the quality of the finished product.
(ii) Quality control sampling and testing results may be used as part of the acceptance decision provided that:

(A) The sampling and testing has been performed by qualified laboratories and qualified sampling and testing personnel.
(B) The quality of the material has been validated by the verification sampling and testing. The verification testing shall be performed on samples that are taken independently of the quality control samples.
(C) The quality control sampling and testing is evaluated by an IA program.
(iii) If the results from the quality control sampling and testing are used in the acceptance program, the SHA shall establish a dispute resolution system. The dispute resolution system shall address the resolution of discrepancies occurring between the verification sampling and testing and the quality control sampling and testing. The dispute resolution system may be administered entirely within the SHA.
(2) The IA program shall evaluate the qualified sampling and testing personnel and the testing equipment. The program shall cover sampling procedures, testing procedures, and testing equipment. Each IA program shall include a schedule of frequency for IA evaluation. The schedule may be established based on either a project basis or a system basis. The frequency can be based on either a unit of production or on a unit of time.

(i) The testing equipment shall be evaluated by using one or more of the following: Calibration checks, split samples, or proficiency samples.
(ii) Testing personnel shall be evaluated by observations and split samples or proficiency samples.
(iii) A prompt comparison and documentation shall be made of test results obtained by the tester being evaluated and the IA tester. The SHA shall develop guidelines including tolerance limits for the comparison of test results.
(iv) If the SHA uses the system approach to the IA program, the SHA shall provide an annual report to the FHWA summarizing the results of the IA program.
(3) The preparation of a materials certification, conforming in substance to Appendix A of this subpart, shall be submitted to the FHWA Division Administrator for each construction project which is subject to FHWA construction oversight activities.
(b) [Reserved].

Sec. 637.209 Laboratory and sampling and testing personnel qualifications.

(a) Laboratories.
(1) After June 29, 2000, all contractor, vendor, and SHA testing used in the acceptance decision shall be performed by qualified laboratories.
(2) After June 30, 1997, each SHA shall have its central laboratory accredited by the AASHTO Accreditation Program or a comparable laboratory accreditation program approved by the FHWA.
(3) After June 29, 2000, any non-SHA designated laboratory which performs IA sampling and testing shall be accredited in the testing to be performed by the AASHTO Accreditation Program or a comparable laboratory accreditation program approved by the FHWA.
(4) After June 29, 2000, any non-SHA laboratory that is used in dispute resolution sampling and testing shall be accredited in the testing to be performed by the AASHTO Accreditation Program or a comparable laboratory accreditation program approved by the FHWA.
(b) Sampling and testing personnel. After June 29, 2000, all sampling and testing data to be used in the acceptance decision or the IA program shall be executed by qualified sampling and testing personnel.
(c) Conflict of interest. In order to avoid an appearance of a conflict of interest, any qualified non-SHA laboratory shall perform only one of the following types of testing on the same project: Verification testing, quality control testing, IA testing, or dispute resolution testing.

Appendix A to Subpart B—Guide Letter of Certification by State Engineer

Date
Project No.

This is to certify that:

The results of the tests used in the acceptance program indicate that the materials incorporated in the construction work, and the construction operations controlled by sampling and testing, were in conformity with the approved plans and specifications. (The following sentence should be added if the IA testing frequencies are based on project quantities. All independent assurance samples and tests are within tolerance limits of the samples and tests that are used in the acceptance program.)

Exceptions to the plans and specifications are explained on the back hereof (or on attached sheet).

Director of SHA Laboratory or other appropriate SHA Official.

[FR Doc. 9515932 Filed 62895: 8:45 am]
BILLING CODE 4910-22-P

APPENDIX B

Summary of Questionnaire Responses

Summary of
Survey Questionnaires for

NCHRP Synthesis Topic 35-01
State Quality Assurance Programs

Agency: ______________________________________

Address: ______________________________________

City: ____________________ State: __________________ Zip: __________

Questionnaire Completed By: ______________________________________

Position/Title: ______________________________________

Date: ____________________

In case of questions please provide:

Telephone: (   ) ________________ Fax: (   ) __________________

Purpose of This Questionnaire

Code of Federal Regulations Part 637 of Title 23 (23 CFR 637) requires each state highway agency to develop a quality assurance (QA) program for the National Highway System (NHS). This program is to assure that the materials and workmanship incorporated into each Federal-aid highway construction project on the NHS are in conformity with the requirements of the approved plans and specifications, including approved changes.

Because the last synthesis on QA was written over 20 years ago and preceded 23 CFR 637, an updated synthesis of practice is needed to describe the states' present QA programs and the future directions for these programs. Since the strategies and practices employed by states to assure quality vary from state to state, it will be very helpful to compile this information in a single document.

Please Return the Completed Questionnaire and Supporting Documents To:

Chuck Hughes
318 Miller School Rd.
Charlottesville, VA 22903
Phone: (434) 823-1797 Fax: (434) 823-1341 e-mail: cshughes3@earthlink.net

Part One Soils and Embankments

1. What type of QA program do you use for Soils and Embankments?
   a. Primarily testing with
      23 agency controlling quality and performing acceptance
      16 contractor controlling quality and agency performing acceptance
      6 contractor controlling quality and contractor results used in the acceptance decision
      1 other (briefly describe). ***On large projects (>100,000 cu. yd) contractor performs quality control and agency does acceptance.***

b. Primarily Methods and Materials provisions where contractor follows prescribed procedures by the agency.
__25__ Yes   *__7__* No

c. Other (briefly describe). ***Pilot project using contractor tests for Acceptance.***

2. What attributes do you require for quality control (QC) and Acceptance
__17__ for QC   *__29__* for Acceptance   Moisture content
__18__ for QC   *__44__* for Acceptance   Compaction
__3__ for QC   *__14__* for Acceptance   Other (briefly describe)
5-Gradation; 4-Atterberg Limits; 3-proof rolling; 2-maximum lab density; 2-R value; 2-AASHTO classification; 2-volume change; 2-% organic content; sand equivalence; % silt; strength, if material is stabilized; thickness of lift; source approval.

3. Using the AASHTO definitions of QC and Acceptance as being two separate functions performed for two different purposes, do you:

a. require the same test methods for QC and Acceptance   *__21__* Yes   *__8__* No

b. require different test methods for QC and Acceptance   *__1__* Yes   *__21__* No

c. specify test methods only for Acceptance   *__23__* Yes   *__14__* No

d. use the same point of sampling for QC and Acceptance   *__13__* Yes   *__15__* No

If No to "d," briefly describe how your agency determines the point of sampling for QC and Acceptance.
4-Random locations for both; 2-contractor decides QC location; 2-Acceptance random; examine for weak conditions; all test points determined randomly by agency; 2-QC at source, Acceptance at site; 8-no QC requirements.

4. For QC for Soils and Embankments do you require:
__19__ an agency established frequency for QC tests
__5__ a contractor established frequency for QC tests
the use of control charts
__19__ none of the above

5. For Acceptance of Soils and Embankments what quality measure(s) do you use?
__3__ Percent within limits (PWL)
Percent defective (PD)
Average absolute deviation (AAD)
Conformal index (CI)
__3__ Average
Standard deviation
__8__ Range
__34__ Individual values
(Other (briefly describe). ______

6. For Acceptance of Soils and Embankments do you use:
__45__ accept/reject/rework. (Briefly describe conditions and amount of pay if less than 100%).
Remove and replace.
pay adjustment system

a. If you use a pay adjustment system do you use:
a stepped pay schedule
an equation; if so, what is the equation:
other (briefly describe). ______

b. For your pay adjustment system do you allow for:
only an incentive
only a disincentive
both

7. Do you use contractor test results as part of the acceptance decision and pay for Soils and Embankments?   *__6__* Yes   *__39__* No
If No, please go to question 8.

a. If Yes, which attributes are used?

6 All
Attributes based on accept/reject
Only attributes that do not involve pay
Other (briefly describe). ______

b. What type of verification system do you use?

4 One contractor test compared to one verification test. Based on:
1 AASHTO D2S tolerance
ASTM D2S tolerance
3 Agency established tolerance
Other (briefly describe basis for the comparison). ______

A comparison of accumulated tests. Based on a statistical comparison of:
F- and *t*-test
only *t*-test
other (briefly describe). ______

5 One agency test compared with several contractor tests.

c. If you use a comparison of accumulated tests is the accumulation based on:

3 lot
day
week
month
1 complete project
production source over extended period
other (briefly describe). ______

d. For verification tests do you take:

3 independent samples
2 split samples
1 both

e. If you use none of the verification systems above, briefly describe how verification is done.

______

8. For training concerning Soils and Embankments, do you require agency and contractor personnel to be trained and/or certified for QC and/or acceptance testing under the following auspices:

a. Agency personnel

In-house program	*32* Training	*22* Certification
Regional programs	*7* Training	*11* Certification
NICET	*1* Training	*1* Certification

4 Other (briefly describe). ***2-University, college; state board of registration; joint agency–industry certification program; WAQTC.***

b. Contractor personnel

Agency in-house program	*12* Training	*12* Certification
Regional programs	*2* Training	*5* Certification
NICET	Training	*1* Certification

3 Other (briefly describe). ***2-University, state board of registration; joint agency–industry certification program; WAQTC (when contractor performs QC).***

Part Two Aggregate Base and Subbase

1. What type of QA program do you use for Aggregate Base and Subbase:
   a. Primarily testing with
      14 agency controlling quality and performing acceptance
      21 contractor controlling quality and agency performing acceptance
      10 contractor controlling quality and contractor results used in the acceptance decision
      other (briefly describe). ______
   b. Primarily Methods and Materials provisions where contractor follows prescribed procedures by the agency.
      15 Yes *14* No
   c. Other (briefly describe). ______

2. What attributes do you require for QC and Acceptance:

27	for QC	*42*	for Acceptance	Gradation
9	for QC	*21*	for Acceptance	Aggregate fractured faces
14	for QC	*24*	for Acceptance	Moisture content
20	for QC	45	for Acceptance	Compaction
	for QC	*3*	for Acceptance	Sand equivalence
	for QC	*5*	for Acceptance	LA abrasion
	for QC	*4*	for Acceptance	Thickness
1	for QC	*12*	for Acceptance	Other (briefly describe). ***4-PI; 3-R value; 2-LL, 2-Washington degradation; width, limerock, shale content; bearing ratio; triaxial shear, wet ball abrasion; thin and elongated particles. (PI = plasticity index; LL = liquid limit.)***

3. Using the AASHTO definitions of QC and Acceptance as being two separate functions performed for two different purposes do you:
   a. require the same test methods for QC and Acceptance *29* Yes *7* No
   b. require different test methods for QC and Acceptance *3* Yes *19* No
   c. specify test methods only for Acceptance *19* Yes *13* No
   d. use the same point of sampling for QC and Acceptance *15* Yes *18* No

   If No to "d," briefly describe how your agency determines the point of sampling for QC and Acceptance. ***8-QC at crusher and 5-Acceptance at road; 4-independent random locations for both; 2-contractor decides QC location; Acceptance at random locations; no QC requirements.***

4. For QC for Aggregate Base and Subbase do you require:
   27 an agency established frequency for QC tests
   8 a contractor established frequency for QC tests
   7 the use of control charts
   18 none of the above

5. What quality measure(s) do you use for acceptance of Aggregate Base and Subbase?
   13 Percent within limits (PWL)
   Percent defective (PD)
   1 Average absolute deviation (AAD)
   1 Conformal index (CI)
   7 Average
   3 Standard deviation
   13 Range
   25 Individual values
   4 Other (briefly describe). ***PWL for CTB; range for UAB; accept/reject at source. (CTB = cement-treated base.)***

6. For Acceptance of Aggregate Base and Subbase do you use:
   35 accept/reject/rework. Briefly describe conditions and amount of pay if less than 100%. ***4-Density failure results in rework; 4-compaction; 2-gradation failure results in price reduction or rejection.***
   16 pay adjustment system. ***4-Applied to gradation.***

a. If you use a pay adjustment system do you use:
   _8_ a stepped pay schedule
   _4_ an equation; if so, what is the equation? ***Gradation % reduction = 5 × % deviation from range limits; price adjustment based on the element.***
   _3_ other (briefly describe). ***Prorated; QLA; QLA with composite PF; actual thickness divided by planned thickness. (QLA = quality level analysis; PF = pay factor.)***

b. For your pay adjustment system do you allow for:
   only an incentive
   _11_ only a disincentive
   _5_ both

7. Do you use contractor test results as part of the acceptance decision and pay for Base and Subbase?
   _13_ Yes   _32_ No
   If No, please go to question 8.
   a. If Yes, which attributes are used?
      _5_ All
      _3_ Attributes based on accept/reject
      _1_ Only attributes that do not involve pay
      _1_ Other (briefly describe). ***Pay reduction for gradation.***

   b. What type of verification system do you use?
      _6_ One contractor test compared to one verification test. Based on:
         _2_ AASHTO D2S tolerance
         ASTM D2S tolerance
         _1_ Agency established tolerance
         _3_ Other (briefly describe basis for the comparison). ***Audits and periodic comparison samples; test witnessing; all tests must pass or go to resolution.***

      _5_ A comparison of accumulated tests. Based on a statistical comparison of:
         _5_ *F*- and *t*-test
         only *t*-test
         other (briefly describe). ______

      _3_ One agency test compared with several contractor tests.

   c. If you use a comparison of accumulated tests is the accumulation based on:
      _6_ lot
      day
      week
      month
      _3_ complete project
      production source over extended period
      other (briefly describe). ______

   d. For verification tests do you take:
      _5_ independent samples
      _2_ split samples
      _3_ both

   e. If you use none of the verification systems above, briefly describe how verification is done.
      ______

8. For training concerning Aggregate Base and Subbase, do you require agency and contractor personnel to be trained and/or certified for QC and/or acceptance testing under the following auspices?

a. Agency personnel

In-house program	***32*** Training	***23*** Certification
Regional programs	***9*** Training	***13*** Certification
NICET	***1*** Training	***1*** Certification

5 Other (briefly describe). ***2-University; 2-joint agency–industry certification program; Crushed Stone Assoc.; state board of registration; certification for production.***

b. Contractor personnel

Agency in-house program	***16*** Training	***15*** Certification
Regional programs	***4*** Training	***8*** Certification
NICET	Training	Certification

4 Other (briefly describe). ***2-University; 2-joint agency–industry certification program; ACI Aggregate Cert.; Crushed Stone Assoc.; state board of registration; certification for production.***

Part Three Hot-Mix Asphalt

1. What type of QA program do you use for Hot-Mix Asphalt:
   a. Primarily testing with
      2 agency controlling quality and performing acceptance
      21 contractor controlling quality and agency performing acceptance
      25 contractor controlling quality and contractor results used in the acceptance decision
      other (briefly describe). ____________________

   b. Primarily Methods and Materials provisions where contractor follows prescribed procedures by the agency.
      2 Yes ***16*** No

   c. Other (briefly describe). ____________________

2. What attributes do you require for QC and Acceptance

43 for QC	***33*** for Acceptance	Gradation
25 for QC	***23*** for Acceptance	Aggregate fractured faces
40 for QC	***40*** for Acceptance	Asphalt content
26 for QC	***23*** for Acceptance	VMA (voids in mineral aggregate)
20 for QC	***26*** for Acceptance	VTM (voids in total mix)
19 for QC	***13*** for Acceptance	VFA (voids filled with asphalt)
13 for QC	***22*** for Acceptance	Thickness
28 for QC	***44*** for Acceptance	Compaction
16 for QC	***39*** for Acceptance	Ride quality
6 for QC	***7*** for Acceptance	Other (briefly describe).

2-Lottman, Hveem Stability; 2-sand equivalence; 2-temperature; 2-BSG; Superpave PG binders; aggregate percent moisture; segregation; TMSG; aggregate friction rating; F/A ratio. (BSG = bulk specific gravity; PG = performance graded; TMSG = theoretical maximum specific gravity; F/A = fines/asphalt.)

3. Using the AASHTO definitions of QC and Acceptance as being two separate functions performed for two different purposes do you:
   a. require the same test methods for QC and Acceptance ***41*** Yes ***3*** No
   b. require different test methods for QC and Acceptance ***1*** Yes ***28*** No
   c. specify test methods only for Acceptance ***10*** Yes ***20*** No
   d. use the same point of sampling for QC and Acceptance ***24*** Yes ***17*** No

   If No to "d," briefly describe how your agency determines the point of sampling for QC and Acceptance. ***7-Independent random locations for both; 4-QC from plant and Acceptance from site; 2-contractor decides QC location; contractor can perform additional tests.***

4. For QC for Hot-Mix Asphalt do you require:
   34 an agency established frequency for QC tests
   12 a contractor established frequency for QC tests
   27 the use of control charts
   4 none of the above

5. What quality measure(s) do you use for acceptance of Hot-Mix Asphalt?
   - *__26__* Percent within limits (PWL)
   - *__1__* Percent defective (PD)
   - *__4__* Average absolute deviation (AAD)
   - Conformal index (CI)
   - *__13__* Average
   - *__3__* Standard deviation
   - *__15__* Range
   - *__4__* Individual values
   - *__3__* Other (briefly describe). ***Warning limits on control charts; moving average.***

6. For Acceptance of Hot-Mix Asphalt do you use:
   - *__6__* accept/reject
   - *__39__* pay adjustment system

   a. If you use a pay adjustment system do you use:
   - *__23__* a stepped pay schedule
   - *__19__* an equation; if so, what is the equation? ***3-PF = 55 + 0.5(PWL); PF = 53 + 0.5(PWL); PF = 83+ 0.2(PWL); 2-multiply individual PFs; various PWL formulas; various equations; total pay = lowest % pay; PF= 100 – 10(ave. deficiency)^ 1.465; equation based on elements; PWL > 93, QAF = 1.05, PWL < 93, QAF = sum PWL for each factor; composite pay factor based on gradation, AC, VTM, density. (QAF = quality analysis factor.)***
   - *__3__* other (briefly describe). ***QLA tables; QLA with composite PF.***

   b. For your pay adjustment system, do you allow for:
   - only an incentive
   - *__7__* only a disincentive
   - *__32__* both

7. Do you use contractor test results as part of the acceptance decision and pay for Hot-Mix Asphalt?
   __29__ Yes *__16__* No
   If No, please go to question 8.

   a. If Yes, which attributes are used:
   - *__20__* All
   - *__3__* Attributes based on accept/reject
   - *__2__* Only attributes that do not involve pay
   - *__5__* Other (briefly describe). ***Less than 100 t/day, contractor tests for mix and agency tests for road; engineer can choose to use or not; pay based on VTM; contractor test results used for VTM and AC. (AC = asphalt concrete.)***

   b. What type of verification system do you use?
   - *__11__* One contractor test compared to one verification test. Based on:
     - *__2__* AASHTO D2S tolerance
     - ASTM D2S tolerance
     - *__9__* Agency established tolerance
     - *__1__* Other (briefly describe basis for the comparison). ***Box sample.***
   - *__10__* A comparison of accumulated tests. Based on a statistical comparison of:
     - *__7__* *F*- and *t*-test
     - *__2__* only *t*-test
     - *__1__* other (briefly describe). ***AAD.***
   - *__14__* One agency test compared with several contractor tests.

c. If you use a comparison of accumulated tests is the accumulation based on:
14 lot
2 day
week
month
3 complete project
1 production source over extended period
2 other (briefly describe). ***Dispute resolution, initially compare first lot (4 sublots), then continue with one sublot per lot for project.***

d. For verification tests do you take:
9 independent samples
9 split samples
7 both

e. If you use none of the verification systems above, briefly describe how verification is done.
____

8. For training concerning Hot-Mix Asphalt, do you require agency and contractor personnel to be trained and/or certified for QC and/or Acceptance testing under the following auspices:

a. Agency personnel

In-house	*30* Training	*26* Certification
Regional programs	*10* Training	*15* Certification
NICET	Training	Certification
NAPA	*1* Training	*1* Certification
Asphalt Institute	Training	Certification

7 Other (briefly describe). ***4-Joint agency–industry certification program; university, college; state board of registration.***

b. Contractor personnel

Agency in-house	*20* Training	*26* Certification
Regional programs	*9* Training	*15* Certification
NICET	Training	Certification
NAPA	*1* Training	*1* Certification
Asphalt Institute	Training	Certification

5 Other (briefly describe). ***3-Joint agency–industry certification program; university, contractor in-house; state board of registration.***

Part Four Paving PCC (three agencies do not use paving PCC)

1. What type of QA program do you use for Paving PCC:

a. Primarily testing with
11 agency controlling quality and performing acceptance
16 contractor controlling quality and agency performing acceptance
13 contractor controlling quality and contractor results used in the acceptance decision
other (briefly describe). ______________________________

b. Primarily Methods and Materials provisions where contractor follows prescribed procedures by the agency.
15 Yes *16* No

c. other (briefly describe). ______________________________

2. What attributes do you require for QC and Acceptance

25 for QC	*26* for Acceptance	Gradation
7 for QC	*6* for Acceptance	Aggregate fractured faces
18 for QC	*31* for Acceptance	Cylinder strength

14 for QC *18* for Acceptance Beam strength
25 for QC *38* for Acceptance Air content
24 for QC *33* for Acceptance Slump
12 for QC *16* for Acceptance Water/cement ratio
2 for QC *3* for Acceptance Permeability
14 for QC *36* for Acceptance Thickness
1 for QC *15* for Acceptance Ride quality
4 for QC *6* for Acceptance Other (briefly describe). ***3-SE; 2-core strength; unit weight; tine texture and surface tolerance; temperature; for QC—dowel bar location, pull test on tie bars, tining depth; aggregate cleanliness; Kelly ball.***

3. Using the AASHTO definitions of QC and Acceptance as being two separate functions performed for two different purposes do you:
   a. require the same test methods for QC and Acceptance *32* Yes *4* No
   b. require different test methods for QC and Acceptance *2* Yes *19* No
   c. specify test methods only for Acceptance *8* Yes *17* No
   d. use the same point of sampling for QC and Acceptance *24* Yes *8* No
   If No, briefly describe how your agency determines the point of sampling for QC and Acceptance.
   4-Acceptance at random locations; 2-contractor decides QC location; 2-QC at source, acceptance at site.

4. For QC for Paving PCC do you require:
   26 an agency established frequency for QC tests
   4 a contractor established frequency for QC tests
   7 the use of control charts
   7 none of the above

5. What quality measure(s) do you use for acceptance of this Paving PCC?
   13 Percent within limits (PWL)
   3 Percent defective (PD)
   Average absolute deviation (AAD)
   Conformal index (CI)
   12 Average
   3 Standard deviation
   15 Range
   10 Individual values
   2 Other (briefly describe). ***Range for non-PWL; PBS model.***

6. For Acceptance of Paving PCC do you use:
   17 accept/reject/rework. Briefly describe conditions and amount of pay if less than 100%. ***Determined by engineer; thickness deficiency > 1 in. R&R; thickness deficiency > 0.6 in. 0 pay; different actions for different attributes; based on compressive strength.***
   28 pay adjustment system. ***2-For deficient thickness.***

   a. If you use a pay adjustment system do you use:
   21 a stepped pay schedule
   7 an equation; if so, what is the equation? ***PWL for thickness plus strength;*** **PF** = ***[0.20(PWL) – 18]/100; based on smoothness and aggregate quality.***
   4 other (briefly describe). ***Tables; % strength deficiency; combination of methods; equation based on elements.***

   b. For your pay adjustment system, do you allow for:
   1 only an incentive
   12 only a disincentive
   16 both

7. Do you use contractor tests results as part of the acceptance decision and pay for Paving PCC?
   14 Yes *25* No
   If No, please go to question 8.

a. If Yes, which attributes are used?
   8 All
   Attributes based on accept/reject
   2 Only attributes that do not involve pay
   3 Other (briefly describe). ***W/C based on contractor batch tickets; engineer chooses to use or not; when contractor elects to use flexural strength option, acceptance is based on flexural strength element.***

b. What type of verification system do you use?
   6 One contractor test compared to one verification test. Based on:
      1 AASHTO D2S tolerance
      2 ASTM D2S tolerance
      2 Agency established tolerance
      2 Other (briefly describe basis for the comparison). ***Engineering judgment; witnessing.***

   4 A comparison of accumulated tests. Based on a statistical comparison of:
      3 *F*- and *t*-test
      1 only *t*-test
      other (briefly describe). ______

   5 One agency test compared with several contractor tests.

c. If you use a comparison of accumulated tests is the accumulation based on:
   4 lot
   2 day
   week
   month
   4 complete project
   3 production source over extended period
   1 other (briefly describe). ***Initially compare first lot (4 sublots), then continue with one sublot per lot for project.***

d. For verification tests do you take:
   5 independent samples
   4 split samples
   4 both

e. If you use none of the verification systems above, briefly describe how verification is done
   ______

8. For training concerning Paving PCC, do you require agency and contractor personnel to be trained and/or certified for QC and/or Acceptance testing under the following auspices:

a. Agency personnel

In-house	*22*	Training	*19*	Certification
Regional programs	*6*	Training	*8*	Certification
NICET		Training		Certification
ACI	*11*	Training	*15*	Certification
PCI	*1*	Training	*1*	Certification

   4 Other (briefly describe). ***University, college; American Concrete Paving Assoc.; state board of registration.***

b. Contractor personnel

Agency in-house	*9*	Training	*13*	Certification
Regional programs	*4*	Training	*6*	Certification
NICET		Training		Certification
ACI	*10*	Training	*15*	Certification
PCI		Training		Certification

   3 Other (briefly describe). ***University, college; American Concrete Paving Assoc.; state board of registration.***

Part Five Structural PCC

1. What type of QA program do you use for Structural PCC?
   a. Primarily testing with
      __14__ agency controlling quality and performing acceptance
      __17__ contractor controlling quality and agency performing acceptance
      __13__ contractor controlling quality and contractor results used in the acceptance decision
      other (briefly describe). ______
   b. Primarily Methods and Materials provisions where contractor follows prescribed procedures by the agency
      __25__ Yes *__9__* No
   c. Other (briefly describe). ______

2. What attributes do you require for QC and Acceptance
   __30__ for QC *__30__* for Acceptance Gradation
   __7__ for QC *__11__* for Acceptance Aggregate fractured faces
   __21__ for QC *__40__* for Acceptance Cylinder strength
   __28__ for QC *__42__* for Acceptance Air content
   __29__ for QC *__40__* for Acceptance Slump
   __15__ for QC *__17__* for Acceptance Water/cement ratio
   __5__ for QC *__8__* for Acceptance Permeability
   for QC for Acceptance Temperature
   for QC *__2__* for Acceptance Beam strength
   __4__ for QC *__10__* for Acceptance Other (briefly describe). ***__2-Temperature; 2-SE; unit weight; soft or friable aggregate particles; surface tolerances; cracking and rebar cover; ride quality on bridge decks; aggregate cleanliness.__***

3. Using the AASHTO definitions of QC and Acceptance as being two separate functions performed for two different purposes, do you:
   a. require the same test methods for QC and Acceptance *__36__* Yes *__5__* No
   b. require different test methods for QC and Acceptance *__1__* Yes *__24__* No
   c. specify test methods only for Acceptance *__10__* Yes *__19__* No
   d. use the same point of sampling for QC and Acceptance *__26__* Yes *__6__* No
   If No to "d," briefly describe how your agency determines the point of sampling for QC and Acceptance. ***__5-Independent random locations for both; QC samples taken from beginning of discharge, acceptance samples taken from middle of discharge.__***

4. For QC for Structural PCC do you require:
   __30__ an agency established frequency for QC tests
   __7__ a contractor established frequency for QC tests
   __5__ the use of control charts
   __6__ none of the above

5. What quality measure do you use for acceptance of Structural PCC?
   __8__ Percent within limits (PWL)
   __2__ Percent defective (PD)
   Average absolute deviation (AAD)
   Conformal index (CI)
   __16__ Average
   __6__ Standard deviation
   __16__ Range
   __8__ Individual values
   __1__ Other (briefly describe). ***__Range for non-PWL.__***

6. For Acceptance of Structural PCC do you use:
   19 accept/reject
   28 pay adjustment system

   a. If you use a pay adjustment system do you use:
      14 a stepped pay schedule
      10 an equation; if so, what is the equation? **PF = [*0.20*(PWL) – *18*]*/100; 1-(strength/f'c); separate equations for strength and permeability.***
      5 other (briefly describe). ***Table, prorated; 4-percent strength deficiency; 2-engineering judgment.***

   b. For your pay adjustment system do you allow:
      only an incentive
      19 only a disincentive
      9 both

7. Do you use contractor tests results as part of the acceptance decision and pay for Structural PCC?
   14 Yes   *28* No
   If No, please go to question 8.

   a. If Yes, which attributes are used:
      10 all
      1 attributes based on accept/reject
      2 only attributes that do not involve pay
      1 other (briefly describe). ***Engineer chooses to use or not use.***

   b. What type of verification system do you use?
      7 One contractor test compared to one verification test. Based on:
         1 AASHTO D2S tolerance
         2 ASTM D2S tolerance
         5 agency established tolerance
         1 other (briefly describe basis for the comparison). ***Witnessing.***

      2 A comparison of accumulated tests. Based on a statistical comparison
         1 *F*- and *t*-test
         1 only *t*-test
         other (briefly describe). ____________

      10 One agency test compared with several contractor tests.

   c. If you use a comparison of accumulated tests is the accumulation based on:
      3 lot
      day
      week
      month
      3 complete project
      2 production source over extended period
      other (briefly describe). ***Initially compare first lot (4 sublots), then continue with one sublot per lot for project.***

   d. For verification tests do you take:
      2 independent samples
      4 split samples
      7 both

   e. If you use none of the verification systems above, briefly describe how verification is done.
      ____

8. For training concerning Structural PCC do you require agency and contractor personnel to be trained and/or certified for QC and/or Acceptance testing under the following auspices:

   a. Agency personnel

In-house	*25* Training	*24* Certification
Regional programs	*6* Training	*10* Certification
NICET	*1* Training	Certification
ACI	*11* Training	*16* Certification
PCI	*1* Training	*1* Certification

   2 Other (briefly describe). ***University, college; American Concrete Paving Assoc.; state board of registration.***

   b. Contractor personnel

Agency in-house	*11* Training	*14* Certification
Regional programs	*4* Training	*7* Certification
NICET	Training	Certification
ACI	*12* Training	*15* Certification
PCI	*1* Training	*1* Certification

   2 Other (briefly describe). ***University, college; American Concrete Paving Assoc.; state board of registration.***

Part Six Independent Assurance

Independent Assurance (IA) is an important function required in a QA program.

1. How is IA organized in your agency?
   - *29* Statewide
   - *16* By district or region
   - *10* By project
   - *10* By system
   - Other (briefly describe). ______

2. To what testing does your agency apply IA inspection/testing?
   - *43* Agency testing
   - *27* Contractor testing
   - *16* Producer testing
   - *10* Supplier testing
   - *3* Consultant testing
   - Other (briefly describe). ______

3. What is the approximate number of full-time equivalent (FTE) IA personnel in your agency? (***See Figures 28 and 29.***)

4. The approximate number of FTE that do sampling and testing? ______

   a. The approximate number of FTE that perform training, record review, etc.?

5. What is the approximate dollar amount of work let to contract last year that was associated with quality assurance?
   ______

Part Seven Use of Consultants and/or Innovative Practices

1. Do you use consultants or other third party to conduct QC or Acceptance tests?
   35 Yes *11* No

   a. If Yes, they are used:
      - *23* in place of agency acceptance tests
      - *20* as a supplement to agency acceptance tests

<u>*12*</u> in place of contractor QC tests
<u>*8*</u> in place of contractor tests used in the acceptance and pay decision
<u>*6*</u> as a supplement to contractor QC tests
<u>*2*</u> as a supplement to contractor tests used in the acceptance and pay decision
other (briefly describe). ____________________

b. Briefly list what products they are used to test. ***13-All; 8-PCC; 5-HMA; 4-prestressed and precast concrete; 3-structural steel and soils; 2-aggregates.***

2. Do you routinely use warranties in your QA program? *8* Yes *37* No
   a. If Yes, are they treated differently than your typical QA program? *4* Yes *4* No
   b. If Yes to "a," briefly describe how they are used. ***2-No agency testing; Materials and Methods warranties used on capital preventive maintenance; seal coats.***

3. Do you routinely use other innovative practices such as design–build, private/public partnership, etc., in your QA program? *9* Yes *36* No
   a. If Yes, briefly describe the products and practices. ***7-Design–build; major bridge replacement and new urban connector road; lane rental; community relations; toll roads; private/public partnership.***

4. Do you accept any pavement materials solely by certification? *11* Yes *34* No
   a. If Yes, briefly list those materials. ***5-Admixtures, 3-reinforcing steel, 2-cement; fly ash, 2-asphalt binders; 2-small quantities and noncritical items; aggregate gradation and cement; curing compound.***

Part Eight Future of QA Program

1. Do you anticipate significant changes in your QA program for any products in the near future?
   26 Yes *19* No
   a. If Yes, which products? Briefly list the products. ***(See Table 11 under Future Programs).***

   b. If Yes, please provide information on the future directions that you anticipate.
   (See Table 12 under Future Programs).

2. Please include any evaluations of QA program effectiveness including reports if they are available.
   Hughes/Killingsworth, Maine DOT Quality Assurance Review, May 2000.

APPENDIX C

Survey Respondents

Alabama
Arizona
Arkansas
California
Colorado
Delaware
District of Columbia
Florida
Georgia
Hawaii
Idaho
Illinois
Indiana
Iowa
Kansas
Kentucky
Louisiana
Maine
Maryland
Massachusetts
Michigan
Minnesota
Mississippi
Montana
Nevada
New Hampshire
New Jersey
New Mexico
New York
North Carolina
North Dakota
Ohio
Oklahoma
Oregon
Pennsylvania
South Carolina
South Dakota
Texas
Utah
Vermont
Virginia
Washington
West Virginia
Wyoming
FHWA Federal Lands Division

Manitoba
Newfoundland
Nova Scotia
Ontario
Quebec

APPENDIX D

Memoranda on Technician and Laboratory Qualification

MEMORANDUM

Subject: **INFORMATION**: Technician Qualification

Date: July 17, 1998

From: Chief, Highway Operations Division

Reply to
Attn of: HNG-23

To: Resource Center Directors
Division Administrators
Acting Federal Lands Highway Program Administrator

The purpose of this memorandum is to serve as a reminder that all sampling and testing of highway materials for Federal-aid projects on the National Highway System (NHS), subsequent to June 29, 2000, must be performed by qualified technicians. Suggestions for what constitutes a technician qualification program are also included.

The regulation for Quality Assurance Procedures for Construction was published as 23 CFR 637 on June 29, 1995. This regulation established a deadline of June 29, 2000, for having all sampling and testing used in the acceptance decision performed by qualified technicians.

The primary objective in establishing technician qualification programs is to assure that the technician is capable of performing the appropriate sampling and testing procedures correctly. In addition, it is likely that a technician will continue to perform the test correctly if they understand the importance of the test and the consequences of conducting improper sampling and testing procedures. The ultimate objective is to assure that maximum quality control and superior highway materials are incorporated into the finished highway infrastructure element.

Technician qualification programs can vary in format while achieving the primary objective of qualified technicians. Currently, several State departments of transportation (DOTs) have combined to develop regional programs that promote reciprocity as well as establishing qualification requirements. Similarly, many individual State DOTs are pursuing their own programs.

While the State or regional flexibility for a technician qualification program format is readily supported, the following items are offered as suggested elements of a complete qualification program:

- Formal training of personnel including all sampling and testing procedures with instructions on the importance of proper procedures and the significance of test results,
- Hands-on training to demonstrate proficiency of all sampling and testing to be performed,
- A period of on-the-job training with a qualified individual to assure familiarity with State DOT procedures,
- A written examination and the demonstration of the various sampling and testing methods,
- Requalification at 2- to 3-year intervals (data from the Independent Assurance program can be used as one element of requalification), and
- The qualification program should have a documented process for removing personnel that perform the sampling and testing procedures incorrectly.

Grandfathering, the acceptance of a Professional Engineer or Engineer-in-Training certificate, or lifetime qualification are not considered to be appropriate criteria for achieving or maintaining qualification status.

Any regulation questions concerning technician qualification should be directed to Jason Dietz, Materials Group, at 202-366-8534. Questions concerning the development of a technician qualification program should be directed to George Jones, Quality Initiatives Group, at 202-366-1554.

/Original Signed by/

Donald P. Steinke

This page last modified on January 23, 2004

MEMORANDUM

Subject:	**INFORMATION**: Technician Qualification	Date:	Oct. 9, 1998
From:	Chief, Highway Operations Division	Reply to Attn of:	HNG-23
To:	Resource Center Directors Division Administrators		

The purpose of this memorandum is to serve as a reminder of some important deadlines and to provide suggestions for what constitutes a qualified laboratory program. It should be noted that 23 CFR 637 does not require all laboratories to be accredited. Only State central laboratories, consultants performing Independent Assurance (IA), and consultants used in dispute resolution need to be accredited.

The regulation for Quality Assurance Procedures for Construction was published as 23 CFR 637 on June 29, 1995. The regulation established three deadlines related to laboratories. The first deadline, June 30, 1997, required each State department of transportation (DOT) central laboratory to become accredited. The second deadline, June 29, 2000, requires non-State DOT laboratories, which perform IA testing or testing for dispute resolution, to become accredited. The third deadline, June 29, 2000, requires all contractors, vendors, and State DOT testing used in the acceptance decision to be performed by qualified laboratories.

The primary objective in establishing laboratory accreditation and qualification requirements is to ensure the capabilities of the laboratories. The ultimate objective is to assure that maximum quality control and superior highway materials are incorporated into the finished highway infrastructure element.

The following should be used as guidance in establishing laboratory qualification programs.

Personnel:

- Supervisors: Supervisors of testing personnel must have a minimum of 3 years experience in testing of highway construction materials.
- Technicians: Technicians must be qualified by a qualification program as was outlined in my July 17, 1998, memorandum on Technician Qualification.

Equipment Documentation:

- A list or record of all laboratory equipment requiring calibration/verification is necessary to provide a qualified laboratory under 23 CFR 637.
- State DOTs should develop test procedures and/or test manuals referencing standard testing procedures, handling, identification, conditioning, storage, retention, and disposal of test samples.

Proficiency in Testing:

- Routinely evaluate testing personnel by observations and split samples or proficiency samples, which will provide the appropriate reviews for field laboratories.

Frequency of Evaluation:

- Laboratory evaluations should be made on a 2- to 3-year cycle.
- Data from the IA program along with observations during IA tests should be used as part of the ongoing evaluation of the laboratory.

Laboratory qualifications that contain all of the above elements will provide the assurance of meeting the objective of testing requirements for NHS projects. Any regulations questions concerning laboratory qualifications should be directed to Jason Dietz, Materials Group, at 202-366-8534.

/Original Signed by/

Donald P. Steinke

This page last modified on January 23, 2004

APPENDIX E

Risks and Operating Characteristics Curves

Understanding risks and how operating characteristics (OC) curves are constructed are important aspects of a quality assurance (QA) program. Selection of acceptance limits relates to the determination of risks. The two types of risk encountered are the seller's (or contractor's) risk, α, and the buyer's (or agency's) risk, β. A well-written QA acceptance plan takes these risks into consideration in a manner that is fair to both the contractor and the agency. Too large a risk for either party undermines credibility. Thus, the risks should be both reasonably balanced and reasonably small. For most highway products, if this is not possible because of the small sample size selected, the risks to the contractor should be less than that to the agency except in critical or life-threatening issues. Risks are defined as:

- Seller's risk (α), also called risk of a type I error—The probability that an acceptance plan will erroneously reject acceptable quality level (AQL) material or construction with respect to a single acceptance quality characteristic. It is the risk the contractor or producer takes in having AQL material or construction rejected.
- Buyer's risk (β), also called risk of a type II error—The probability that an acceptance plan will erroneously fully accept (100% or greater) rejectable quality level (RQL) material or construction with respect to a single acceptance quality characteristic. It is the risk the highway agency takes in having RQL material or construction fully accepted. [The probability of having RQL material or construction accepted (at any pay) may be considerably greater than the buyer's risk.]

The α and β risk levels that might be appropriate vary depending on the material or construction process that is involved. The appropriate risk level is a subjective decision that can vary from agency to agency. In reality, it is likely that few agencies have developed and evaluated the risk levels associated with their acceptance plans.

The concept of α and β risks derives from statistical hypothesis testing where there is either a right or wrong decision. As such, when α and β risks are applied to materials or construction they are only truly appropriate for the case of a pass/fail or accept/reject decision and, indeed, may lead to considerable confusion if an attempt is made to apply them to the pay adjustment case. When materials not only can be accepted or rejected, but can also be accepted at an adjusted pay, then additional interpretations or clarifications must be applied to the definitions of risks (*12,20*).

For example, in the previously mentioned definition for buyer's risk, it states that β is the probability that RQL material may be accepted at 100% pay or greater. The definition must then go on to point out that there is also a probability that the RQL material will receive some reduced pay. Although it is not stated as directly, the same reasoning is true for the seller's risk. The definition indicates that α is the probability that AQL material will be rejected. Although not stated in the definition, it is also true that there is a probability that the AQL material will be accepted at a reduced pay (*12*).

Operating characteristic curves—α and β are very narrowly defined to occur at only two specific quality levels. β is the probability of accepting (at full pay or more) material that is exactly at the RQL level of quality, whereas α is the probability of rejecting material that is exactly at the AQL level of quality. These definitions do not, therefore, provide an indication of the risks over a wide range of possible quality levels. To evaluate how the acceptance plan will actually perform in practice, it is necessary to construct an operating characteristic (OC) curve.

Definition of OC curve—A graphic representation of an acceptance plan that shows the relationship between the actual quality of a lot and either (1) the probability of its acceptance (for accept/reject acceptance plans) or (2) the probability of its acceptance at various pay levels (for acceptance plans that include pay adjustment provisions) (*13*).

Example of OC curve—An example of an OC curve for a pass/fail or accept/reject acceptance plan, case (1) in the previous definition, is shown in Figure E1. Probability of acceptance is shown on the vertical axis for the range of quality levels indicated on the horizontal axis. An example of an OC curve for an acceptance plan with pay adjustment provisions, case (2) in the previous definition, is shown in Figure E2. The axes are the same as for Figure E1, but there are multiple curves, one plotted for each of several selected pay levels.

Each curve plotted in Figure E2 represents the probability of receiving a pay factor equal to or greater than the one indicated for the line. For example, for the OC curves in Figure E2, material that is of exactly AQL quality has approximately a 45% chance of receiving a pay factor of 1.04 (104%) or greater. This same AQL material has approximately a 60% chance of receiving full pay (100%) or greater, which also means that it has approximately a 40% chance of receiving less than 100% pay (*20*).

On the other hand, for the OC curves in Figure E2, material that is of exactly RQL quality has approximately a 50% chance of receiving a pay factor of 0.80 (80%) or greater, and approximately an 80% chance of receiving a pay factor of

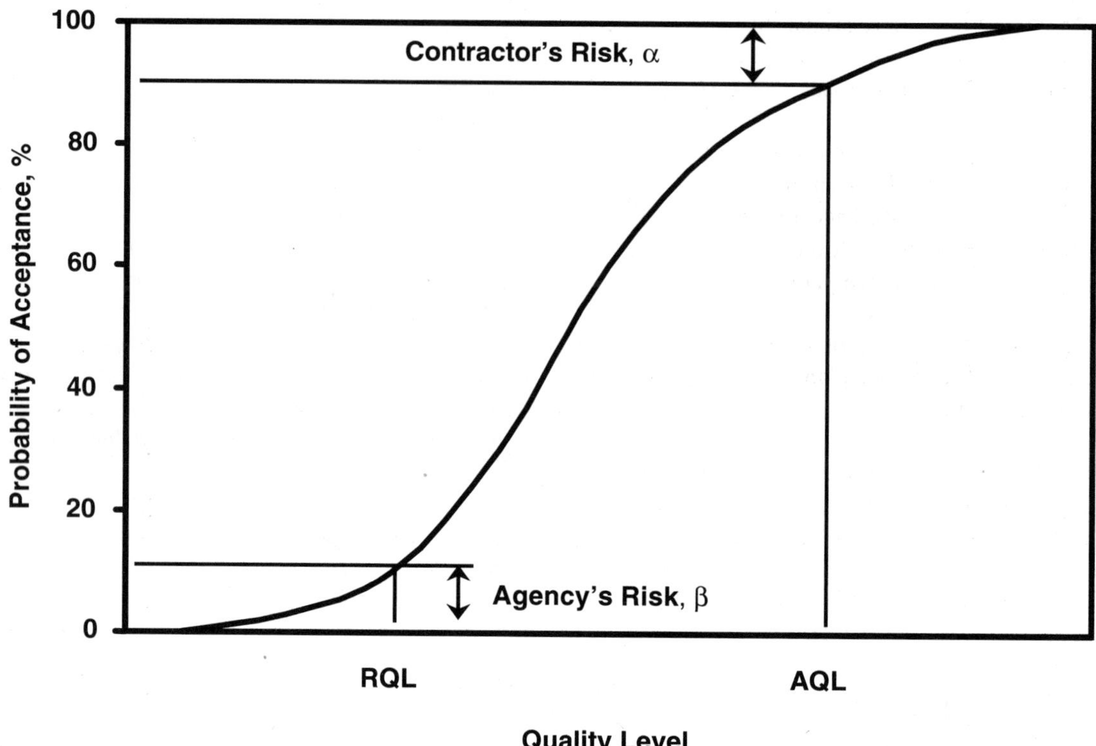

FIGURE E1 Typical OC curve for an accept/reject acceptance plan (*20*).

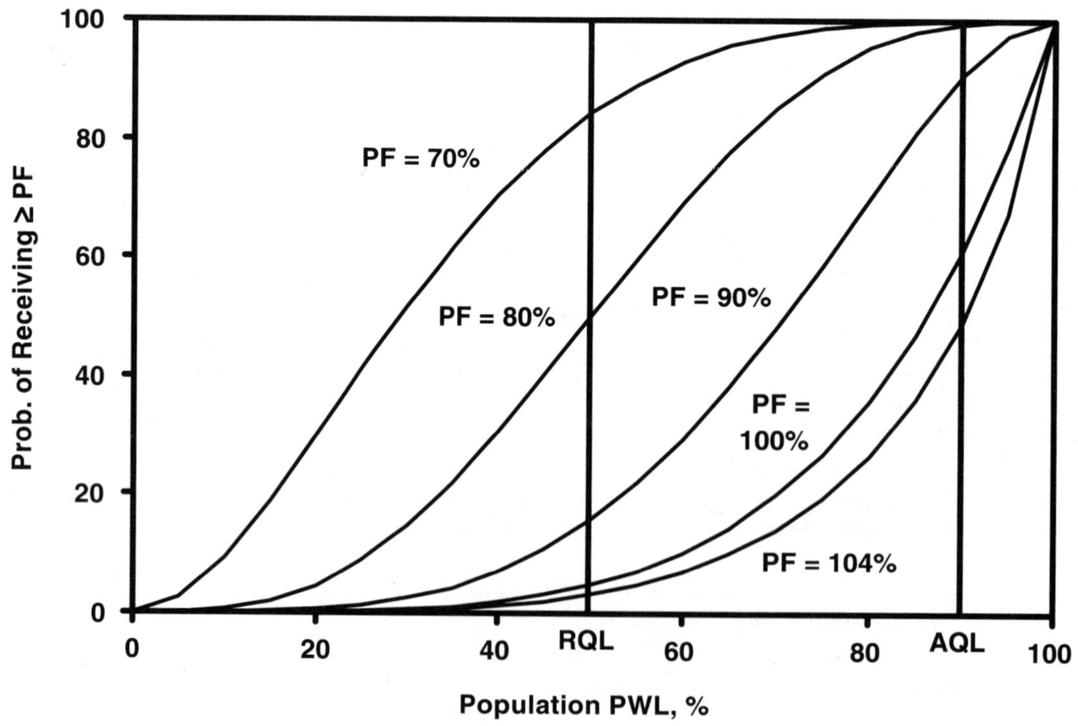

FIGURE E2 Typical OC curves for an acceptance plan with pay adjustments (*20*).

0.70 (70%) or greater. Similar pay probabilities can be determined for any level of actual quality, and additional curves could be developed for any specific pay factor.

Pay adjustment system plans—As discussed earlier and shown in Figure E2, the consideration of only α and β risks is not sufficient when pay adjustments are used. From Figure E2 it can also be seen that using multiple OC curves is not an easy way to evaluate an acceptance plan.

Expected pay curves—Thus, another way to present the pay performance for an acceptance plan is with what is called an expected pay (EP) curve.

Definition of EP curve—A graphic representation of an acceptance plan that shows the relation between the actual quality of a lot and its EP; that is, mathematical pay expectation, or the average pay the contractor can expect to receive over the long run for submitted lots of a given quality (*13*). Both OC and EP curves should be used to evaluate how well a pay adjustment acceptance plan is theoretically expected to work.

Example of EP curve—An example of an EP curve is shown in Figure E3. Quality levels are indicated on the horizontal axis in the usual manner, but instead of probability of acceptance, the vertical axis gives the expected (long-term average) pay factor as a percent of the contract price.

Although the risks have a different interpretation when associated with EP curves than with OC curves, the same type of information is provided. It is a generally accepted tenant that the average pay for material that is just fully acceptable should be approximately 100% of the contract price. For the example in Figure E3, AQL work receives an expected pay of 100%, as desired, whereas truly superior work that is better than the AQL receives an expected pay of 102%. At the other extreme, RQL work corresponds to an expected pay of 70%. For still lower levels of quality, the curve stops at a minimum expected pay of 50%.

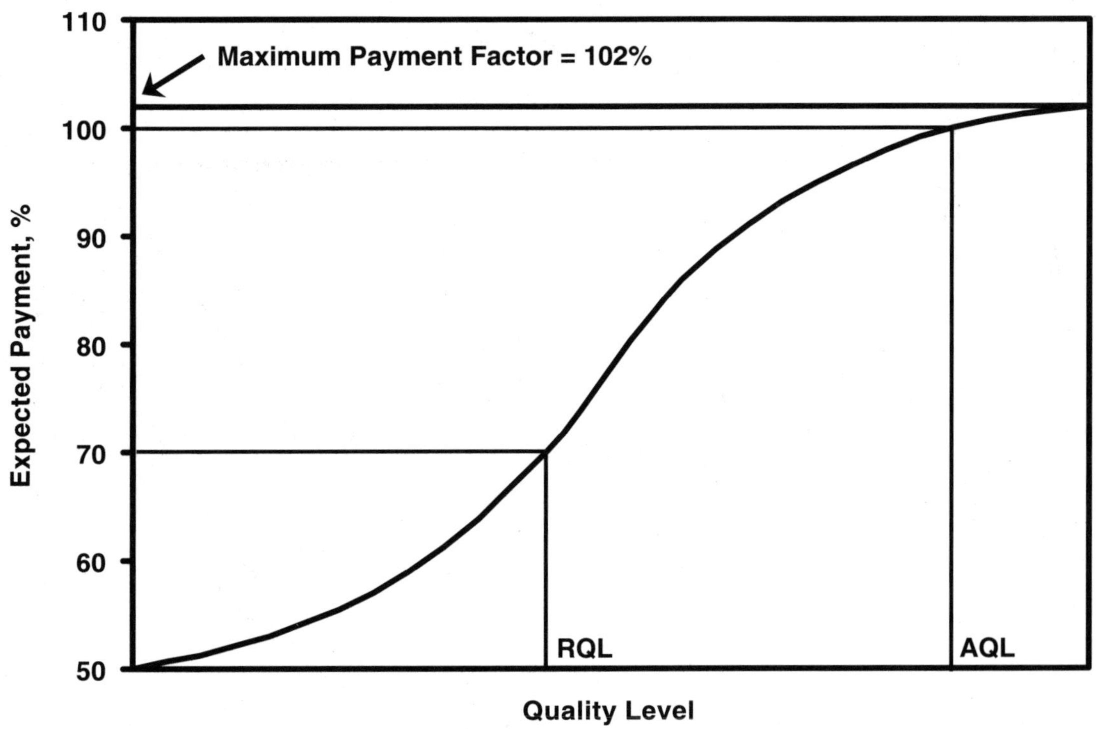

FIGURE E3 Typical expected payment curve (*20*).